Louis G. Nickell

Plant Growth Regulators

Agricultural Uses

With 29 Figures

Springer-Verlag
Berlin Heidelberg New York 1982

Louis G. Nickell
Research and Development, Velsicol Chemical Corporation
341 East Ohio Street
Chicago, Ill. 60611/USA

ISBN-13: 978-3-642-68206-3 e-ISBN-13: 978-3-642-68204-9
DOI: 10.1007/978-3-642-68204-9

Library of Congress Cataloging in Publication Data. Nickell, Louis G., 1921– Plant growth regulators. Bibliography: p. Includes index. 1. Plant regulators. I. Title. SB128.N52.631.8.81-9320

Softcover reprint of the hardcover 1st edition 1982

2152/3140-543210

To my wife Natalie

lifelong companion, closest confidant, strongest supporter

Preface

Chemicals that control plant growth have long been treated like a poor relation of the herbicides yet in one manner of thinking, herbicides themselves are but one facet of the entire picture of plant growth regulation – a major fraction, to be sure, economically. It is now time to recognize that plant growth regulators should occupy an increasingly important role in agriculture. Sufficient numbers of uses having considerable economic return have already become established:

(a) to increase the latex flow in the rubber trees;
(b) to ripen sugarcane;
(c) to control sprouting in onions and potatoes;
(d) to shorten and strengthen wheat stems to prevent lodging;
(e) to prevent premature deterioration; and
(f) to permit control of timing for maximum utilization of crops.

In addition, as energy becomes more difficult and costly to obtain, plant growth regulators will play an increasingly important role in energy conservation as a result of increased yields due to their use.

There are a number of ways to present to the reader the role and effectiveness of plant growth regulators. The one chosen here is to emphasize the effects on plant functions such as

the induction of roots,
the control of flowering,
the control of sex, and
the control of aging.

Little emphasis has been placed on the basic research that has served as a background for the successes and potential successes discussed herein. Nor is much attention paid to the mode of action of the various regulators. Instead, each section includes an extensive bibliography that includes numerous review articles written by investigators particularly skilled and familiar with special aspects of research concerned with the regulation of plant growth. The emphasis has been placed on the successes already achieved in the use of plant growth regulators for a number of purposes. Also included is information concerning other materials that have been shown to be active but not yet established in a commercial sense. No attempt has been made to concentrate on the acceptability of the use of the various compounds by regulatory agencies throughout the world; such could be a never-ending task and also subject to so many changes that it could never be up-to-date. Mention of registration and acceptability is made only to point out the commercial successes enjoyed through the use of certain compounds for specific uses.

The first time a compound is mentioned, the chemical name is given, followed by the common name or abbreviation. To the greatest degree possible, common names for chemicals have been used. Also from time to time trade names and/or code designations are used when it is felt they are more appropriate. The relationship and interchangeability of the chemical name, the common name, the trade names and, in some instances, the code numbers of designations will be found throughout the text in various tables and are summarized in the large table following the last chapter. Producers or suppliers are not necessarily emphasized but are sometimes mentioned to show where the emphasis is being placed for the search in a given area of plant growth regulation.

Instructions on the details of how to use the various growth regulators discussed is purposely avoided. In most instances, concentrations of chemicals and/or amounts to be used are also not stressed for a host of reasons – including the great variability in dosage according to climatic conditions, time of year, age of crop, geographic location, and specific results desired. For such information the reader is referred to the extensive bibliography and to the labels prepared by the producing companies.

Originally it was proposed to include a chapter on the screening methods for determining and evaluating plant growth regulators. This plan was abandoned because of the high degree of specificity when dealing with the regulation of crop growth. A biologically active compound might well have some activity on a wide range of commercial crops and yet have the desired or practical effect on only one or a very few. Because of this, there is an increasing tendency not only to develop a test for screening for a given type of activity but to apply that test to the specific crop concerned. There are many examples, such as the differential effect of a regulator on apples versus its action on peaches; a special example appears to be the unique sugar enhancing activity of vanillin on sugarcane and grapes. It is felt that the reader can obtain a clear picture of how the activity of a given growth regulator was determined and evaluated by reading the literature cited for a particular crop and for the specific plant process affected by the regulator.

The real purpose of this book is (a) to show the tremendous advancements made in the use of plant growth regulators in agriculture and in horticulture during the last several decades, (b) to present a comprehensive study of the literature, and (c) to summarize the author's personal knowledge of the status, both scientific and commercial, of this field as of mid-1980.

A symposium on herbicide antidotes was held in 1977 at the national meeting of the American Chemical Society in New Orleans, Louisiana. In the preface to the published proceedings of this symposium, the editors quoted a statement made in 1974 by Andre and Jean Mayer (Daedalus 103,

83, 1974). It is most appropriate that this quotation be repeated in the introduction to the present volume:

"Few scientists think of agriculture as the chief, or model, science. Many, indeed, do not consider it a science at all. Yet it was the first science – the mother of sciences; it remains the science that makes human life possible; and it may well be that, before the century is over, the success or failure of science as a whole will be judged by the success or failure of agriculture."

November 1981 L. G. Nickell

Table of Contents

Chapter 1 Introduction 1

Chapter 2 Rooting and Plant Propagation 4

Chapter 3 Germination and Dormancy 6

Chapter 4 Flowering 8

Chapter 5 Gametocides 15

Chapter 6 Abscission 19

Chapter 7 Fruit Set and Development 28

Chapter 8 Plant and Organ Size 32

Chapter 9 Axillary Buds 45

Chapter 10 Chemical Pruning 47

Chapter 11 Plant Shape 49

Chapter 12 Tillering 50

Chapter 13 Resistance to, and Control of, Insects and Diseases . . 51

Chapter 14 Overcoming Environmental Stress 55

Chapter 15 Mineral Uptake 59

Chapter 16 Plant Composition 60

Chapter 17 Metabolic Effects, Ripening, and Yield Increases . . . 63

Chapter 18 Modification of Sexual Expression 89

Chapter 19 Senescence 91

Chapter 20 Desiccation 93

Chapter 21 Protection Against Herbicide Damage 99

Chapter 22 Increase of Herbicide Absorption and Translocation . 103

Chapter 23 Toxicology, Environmental and Human Safety 104

Chapter 24 Summary 107

Literature 126

Author Index 150

Subject Index 161

Chapter 1. Introduction

It has been established, and is now well accepted, that normal plant growth and development is controlled by chemicals produced by the plant itself (endogenous plant hormones). It is also apparent that synthetic growth-regulating chemicals are becoming extremely important and valuable in the commercial control of crop growth, in both agriculture and in horticulture [1]. The general thinking is that synthetic regulators produce their effects through changing the endogenous level(s) of the naturally-occurring hormones – allowing a modification of growth and development in the desired direction and to the desired extent. Much attention has been given in the popular press to the potential for use of plant growth regulators in controlling crop plants as well as horticultural plants; the predictions for success range from a reluctant conservatism to unlimited enthusiasm. Although use of plant growth regulators is presently small in volume and value compared to their more important pesticide relatives – herbicides, insecticides, and fungicides – predictions are that the rate of dollar volume will increase much faster for plant growth regulators than for pesticides during the next several years, thus being the most rapidly expanding segment of the agricultural chemical business.

Plant growth regulators usually are defined as organic compounds, other than nutrients, that affect the physiological processess of growth and development in plants when applied in low concentrations. For practical purposes, plant growth regulators can be defined as *either natural or synthetic compounds that are applied directly to a target plant to alter its life processes or its structure to improve quality, increase yields, or facilitate harvesting.*

By this latter definition, pesticides, when applied to produce specific beneficial changes in the target crop, also can be considered plant growth regulators. The term "plant hormone", when correctly used, is restricted to naturally-occurring plant substances; these fall into five classes: auxins, gibberellins, cytokinins, inhibitors, and the gas ethylene. The term "plant growth regulator" is not restricted to synthetic compounds but also includes the naturally-occurring hormones. Thus, the term "regulator" is very broad and is the one that will be used in this discussion.

Auxins are compounds that cause enlargement of plant cells.

Gibberellins are compounds that stimulate cell division, cell enlargement, or both.

Cytokinins stimulate cell division in plants.

Inhibitors are a diverse group of plant hormones that inhibit or retard a physiological or biochemical process in plants.

The specific actions of these classes of hormones are those primarily connected with each type; they are but a few of the known effects of those classes of substances. Plant scientists are divided over the relative importance and position of the gas ethylene in the hierarchy of plant growth substances. It plays an important but as yet elusive role in plant bioregulation. This ubiquitous compound is constantly being generated in plants and is continually escaping from them as a gas. It does not seem to enter into chemical reaction with plant parts, and yet to exert its influence it must be involved in some such way. Ethylene influences a wide variety of plant processes even in extremely small amounts. Since the early 1960's, the importance of ethylene has been recognized more and more. To emphasize the importance of ethylene, many investigators believe that both natural and synthetic plant growth regulators may exert their influences through their effect on ethylene production and/or activity.

A discussion of plant growth regulators can be divided in a number of ways: by crop, by chemical group, by physiological process, and by miscellaneous classifications that do not easily fit into any other easily dividable classifications. The use of categories by crops would be rather confining because practical and government-approved uses for plant growth regulators exist for only a few crops; consequently, the breakdown would probably be by process within each category of crop. The division of discussion by chemical class would be equally confusing because some chemicals have very restricted uses on a few crops, others have restricted uses on one crop, and a few have multiple uses on a number of crops. Thus, division by chemicals would result in a skewing of the discussion, probably limiting it to a half-dozen compounds or classes. The discussion in this book will be primarily by plant process, with some sections, such as insect control, that do not fall into an obvious plant process. In using this approach, it should make it easier to bring to the attention of the reader (a) the practical and approved uses of plant growth regulators, and (b) the stage and development of those chemicals that are still under preliminary or advanced evaluation. In some chapters the division by plant process is not clear-cut because of the paucity of available practical information, thus making it necessary to combine two or more related topics.

The response of a plant or a plant part to a plant growth regulator may vary with the variety of plant. Even a single variety may respond differently, depending on its age, environmental conditions, physiological state of development (especially its natural hormonal content), and state of nutrition. Thus, whenever a general rule is suggested concerning the action of a specific growth regulator on plants, exceptions almost always can be found. There are several proposed modes of action for each class of plant hormone, with substantial arguments for and against each mode. The total results suggest a dual effect of plant hormones on plant growth: (a) a rapid growth response brought about by modification of system(s) present in the cell membrane at the time of application, and (b) a delayed growth response occurring later by action on enzymes formed after application. The primary site of action of plant growth hormones at the molecular level remains unresolved. As the dean of American plant physiologists, Dr. Kenneth V. Thimann, stated in his recent review of 50 years of plant hormone research, "I am afraid there will have to be some more theories before a definite mode of action for any plant hormone is found" [2]. There are two main reasons why the mechanisms of plant hormone action are still obscure: first, each hormone produces a great variety of physiological responses; second, several of these responses to different hormones frequently are similar.

The first clear indication that a growth hormone occurs naturally in plants was the demonstration by Went in 1926 [3] that oat seedlings contained a diffusible substance that would promote their growth. The discovery eight years later by Kögl and co-workers [4] that *indoleacetic acid* (IAA) is capable of promoting the elongation of plant cells focused considerable attention on this compound, now recognized as the most important of the auxins.

Since the first practical uses of plant growth regulators – stimulating root formation, and promoting flowering in pineapple by applying acetylene and ethylene [5, 6, 7] – the importance of such chemicals in agriculture has steadily increased. Although this increase was slow for many years, a marked change occurred in the 1970's. New compounds and new uses are constantly being reported in the scientific literature, and particularly important is the heart-warming decrease in the time between scientific discovery and practical utilization. In his review of plant growth regulators in 1966, John W. Mitchell of the U. S. Department of Agriculture predicted, "New ways of advantageously controlling plant growth and behavior with hormones and growth regulators are sure to be discovered" [8]. The intervening fifteen years have shown him to be correct. Moreover, his statement can be repeated today and surely will be equally effective fifteen years in the future.

What is the future for plant growth regulators? An ever-increasing role for growth regulators is assured by the certain increase in the cost of energy, the continual decrease in prime productive land as it is converted to urban and industrial uses, and the certain need to double the world's food production by the end of the century. The diversity of the effects of growth regulators is outstanding; there seems to be virtually no limit to the number of ways in which growth regulators may be used. These ways range from the more obvious effects such as change in shape, size, or growth rate, to bringing about more subtle changes in metabolism that affect both the quality and quantity of the desired product. The need to increase agricultural production at an unprecedented rate almost dictates that plant growth regulators will be a major contributor in reaching the desired goals.

Chapter 2. Rooting and Plant Propagation

One of the oldest uses for plant growth regulators has been to initiate and/ or accelerate the rooting of cuttings [9]. Probably the best and most commonly used chemical for this purpose is *indolebutyric acid* (IBA), which is decomposed relatively slowly by the auxin-destroying enzyme systems in plants. Because this compound also moves very slowly in the plant, much of it is retained near the site of application – another desirable characteristic. Another highly active auxin frequently used for root promotion is *naphthalene acetic acid* (NAA). As NAA is more toxic than IBA, there is a greater danger of injury to treated plants. The amides of both compounds are also effective rooting agents. Because the amide of NAA is less toxic than the acid itself, it is safer to use. Many phenoxy compounds, including *2,4-dichlorophenoxyacetic acid* (2,4-D) and *2,4,5-trichlorophenoxyacetic acid* (2,4,5-T), promote root formation if used in low concentrations; because their toxicity limit is near the optimal concentration for initiation of roots, the effective range of these compounds is quite narrow. The type of root systems produced varies with the growth regulator used. The phenoxy acids tend to produce bushy, stunted, and thickened root systems, whereas IBA produces a strong fibrous root system.

Despite the many other chemicals that have been evaluated for their effects on rooting – and even though some of the them have desirable actions – IBA remains the compound of choice. Since the work of Zimmerman, Hitchcock and others in the 1940's showing the root-inducing action of many synthetic and naturally-occurring auxins, this use of IBA has become routine throughout the world [10–15]. More recent publications have shown the effectiveness of IBA on a practical scale throughout the world with

rose [16, 17], tea [18], winged bean [19], rhododendron [20, 21], grapes [22, 23], *Peperomia* [24], *Protea* [25, 26, 27], *Ficus* [28], *Ixora* [29, 30], bougainvillea [30], gardenia [30], jasmine [31], eggplant [32], hibiscus [30], *Mussaenda* [30], and *Nyctanthes* [30], as well as many bearing and forest trees [33], including apples [34], peach [35, 36], plum [37], carob [38], poplar [39], pine [40, 41], rubber [42], arborvitae [39], *Cryptomeria* [39, 43], *Cydonia* [36], mahogany [44], yemane [44, 45], and tulip tree [46].

There are three major methods for applying growth regulators to stem cuttings for the induction of roots: the quick dip method, the prolonged soaking method, and the powder method. In the quick dip method the basal ends of cuttings are dipped for a few seconds in a concentrated solution of the growth regulator in alcohol (up to 10,000 ppm). In the prolonged soaking method, the basal end of cuttings are soaked in dilute solutions (10 to 500 ppm) for up to 24 hours. In the powder method, the basal end of cuttings are treated with the growth regulator in a carrier – usually a clay or a talc. The concentration of the active ingredient in the inert carrier is between the two other methods (i.e., 500 to 1000 ppm).

During the past decade, a considerable number of articles concerning the effect of abscisic acid on rooting have been published, with mixed results. Promotion of rooting of cuttings has been reported for runner beans [47], mung bean [48, 49], pea [50], ivy [48], tomato [51], grape [52], and poplar [53]. No response has been reported for poinsettia [54], cantaloupe [55], common bean [56], and others [56, 57], while rooting was inhibited in *Zebrina pendula* [58]. Many of these studies were conducted to determine the role of growth regulators in the induction of rooting. It is hard to imagine that abscisic acid will replace IBA in commercial usage for this purpose.

Chapter 3. Germination and Dormancy

Germination

In many cases, seed germination can be promoted or inhibited by application of naturally-occurring plant growth regulators [59]. Although inhibitors, such as abscisic acid, retard germination in most species not all seeds can be released from dormancy by growth promoters. Of the natural hormones, gibberellic acid is the most potent germination promoter, breaking seed dormancy in a wide range of species: peach [60], wild mustard [61], stinkweed [61], wild oats [62], citrus (sour orange [63], rough lemon [63], trifoliate orange [63], and sweet orange [64]), celery [65, 66, 67], sorghum [68], cotton [69, 70], beans [71], peas [71], lettuce [72], Chinese cabbage [73], cucumber [74], and other cucurbits, *Impatiens balsamina* [75], *Lavandula angustifolia* [75], *Brassica rapa* [75], and *Viola odorata* [75]. Light-sensitive lettuce seed can germinate in darkness if treated with gibberellic acid [76, 77]. With respect to the number of species affected, the range for the cytokinins and ethylene in promoting germination is more restricted. Miller showed that seeds of certain lettuce varieties (that require light for germination) will germinate in the dark if first treated with kinetin [78]. Auxins are even less effective in their ability to promote germination; although they do promote germination slightly in some cases at low concentrations [79], generally they are ineffective – and sometimes inhibitory [80, 81, 82].

The synthetic growth retardants chlormequat, daminozide and phosphon D inhibit the germination of lettuce seed [83]. On the other hand, the fungal metabolite, fusicoccin, is an extremely potent general germination promoter [74, 84, 85].

Dormancy

Although the dictionary defines dormancy as "having biological activity suspended", in fact it is a difficult term to define. Numerous definitions have been given [86]. Dormancy is induced by two general types of conditions: a) external, such as unfavorable environmental conditions; and b) internal, which prevent growth even though the external conditions are favorable. These two types of dormancy are given the terms *quiescence* and *rest*, respectively [87].

Dormancy occurs in many plant parts including buds, bulbs, and seeds. Seed dormancy can be due to any of a number of factors, or a combination of such factors: (a) rudimentary embryos, (b) physiologically immature embryos, (c) mechanically resistant seed coats, (d) impermeable seed coats, and (d) presence of germination inhibitors [88].

Seeds of many plants require a period of after-ripening at low temperatures before they will germinate. Gibberellins have been shown to eliminate this cold requirement in a number of plants and to shorten the period required in others.

Bud rest in many plants can be terminated or prolonged by the appropriate application of plant growth regulators. Probably one of the best known uses of plant growth regulation is the suppression of sprouting in potatoes and onions by treatment with maleic hydrazide [89–92]. On the other hand, the ability of gibberellins to break the dormancy of potato tubers was demonstrated as soon as commercial amounts became available [93, 94, 95].

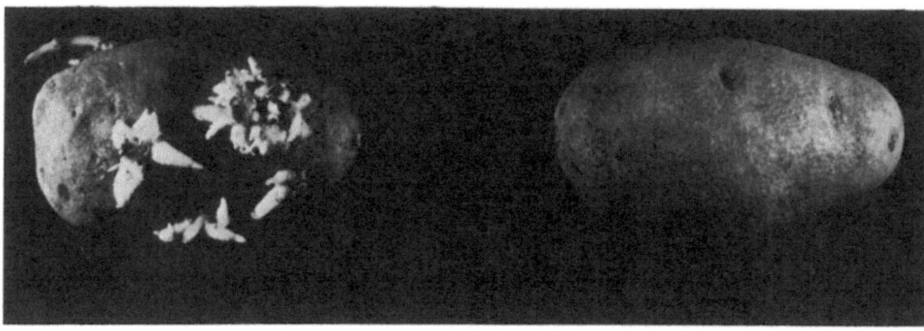

Untreated Treated

Fig. 3-1. Suppression of sprouting in potatoes by treatment with maleic hydrazide. Photograph courtesy of Uniroyal Chemical

Fig. 3-2. Suppression of sprouting in onions by treatment with maleic hydrazide. Left: untreated; right: treated. Photograph courtesy of Uniroyal Chemical

Chapter 4. Flowering

Control of lowering is one of the most important practical aspects of horticulture and agriculture. With many horticultural crops, the key to financial success is the capability to induce flowering and more importantly, to induce it on command in order to meet certain major market and holiday dates. Conversely, the ability to prevent flowering is extremely important in agricultural crops when flowering often causes a decreased economic benefit.

It is said that the use of smoke to bring about floral initiation in pineapples was discovered in 1893 when a carpenter working in a greenhouse in the Azores accidentally set fire to a pile of shavings; to the surprise of the grower, the plants burst into flower instead of being damaged. By the 1920's it was a recognized fact that pineapple could be forced to flower by smoke from fires (used during cold weather to prevent stoppage of growth) and that this effect was caused by the smoke's content of unsaturated gasses such as ethylene [96]. By the early 1930's ethylene was shown to accelerate flowering in pineapple, and soon acetylene gas was used commercially in Hawaii to force the initiation of flowering. In the 1940's auxins were shown to produce this effect, and naphthaleneacetic acid was the next forcing agent to be used commercially on pineapples.

Although hydrazines are more commonly thought of as growth retardants, several of them, particularly beta-hydroxyethyl hydrazine, were shown in the mid 1950's to induce flowering in Hawaiian pineapples [97]. Abeles [98] points out that all those compounds that promote flowering in pineapples and other bromeliads: (a) are ethylene analogs such as acetylene; (b) produce ethylene chemically (ethephon and beta-hydroxyethyl hydrazine), or (c) cause the plants to evolve additional quantities of ethylene themselves (naphthaleneacetic acid and 2,4-D). The latest group of compounds demonstrated to be effective forcing agents for pineapple are the haloethanephosphonic acids such as ethephon; ethephon at 1 to 2 lb/acre produces total floral induction [99].

Several theories have been advanced to explain the mechanism of floral induction in pineapples. One is that ethylene causes a sensitivity to naturally-occurring auxins in the growing vegetative tip. Another is that exogenously applied auxin lowers the level of naturally-occurring auxins to that which is best for inducing flowers. Still another is that flowering is initiated by the accumulation of auxins in the vegetative tip; because ethylene treatment actually does not increase auxin content but does induce flowering, most investigators doubt this theory.

In commercial pineapple cultivation, uniform ripening of fruits is essential. The desired uniformity of flowering is now assured by the use of plant growth regulators [100, 101] and is done commercially in Hawaii [97, 102], Florida [103], Puerto Rico [104, 105], Kenya [106], Australia, India [107], Sri Lanka, Mexico, Malaysia [108], South Africa [109], the Philippines, Taiwan [110], and most recently in Ghana [111, 112, 113].

The vegetative phase of bud development that precedes flower initiation is critically important in determining the amount of blossom produced. In 1975, Luckwill [114] suggested that growth regulators which promote or inhibit flowering in apple could exert their effects indirectly by modifying the rate of node production as opposed to directly upon the floral initiation process itself. He believed that this might explain why gibberellins, which in some plants promote flowering, are potent inhibitors of flowering in apples and many other spring flowering trees and shrubs that form their floral initials in the previous summer. Growth retardants such as daminozide, believed to act by blocking the biosynthesis of endogenous gibberellins, have the opposite effect of promoting flowering in these species. Chemicals that delay bud opening offer the possibility of reducing spring freeze injury in tree crops [115–117]; a number of chemicals have been tested for this purpose with mixed results. Autumn sprays of gibberellic acid have retarded bloom in grapes [118], stone fruits [119–123], and almonds [124]. Autumn sprays with daminozide delay the initiation of bloom in pears [125], apples [126, 127], and grapes [128, 129].

Many annual vegetables, such as lettuce, radish, mustard, and dill, normally flower only when days are long but can be made to flower early by treatment with gibberellins. Many biennial vegetables, such as carrots, beets, and cabbage, require low temperatures to flower but these vegetables also flower after treatment with gibberellins.

To improve yields, it is commercially worthwhile to prevent flowering in some crops such as sugarcane. In other crops, among them almond, peach, and tall oil trees, delaying the onset of flowering may be useful to avoid adverse weather conditions such as extremes in temperature and moisture. Furthermore, such a delay (a) can bring two plant varieties with different flowering dates into synchronization for breeding purposes (as in the case of varieties of almond trees), or (b) can control the timing of flowering plants such as carnation or poinsettia to coincide with major holidays when selling prices are higher [1].

Flowering in plants is the end product of the cumulative effect of many subtle metabolic changes resulting in the initiation of flower buds. In 1920, Gardner and Allard [130] established that many plants show a peculiar sensitivity to light in that flower buds will be initiated in such plants only under certain day lengths. Subsequent work established that the night length is a critical period. Sugarcane is considered to be among the most sensitive of plants to light. Sugarcane was classified by Allard in 1938 as an intermediate type but later work showed that sugarcane is a short day plant, initiating flowers only within a critical range of day lengths. Members of this class must have an uninterrupted dark period in order to flower. The briefest interruption, for sugarcane as little as 50 footcandle-minutes [1, 131, 132] of incandescent light is usually sufficient to prevent flowering.

After it had been determined that night interruption from September 1 to 20 would inhibit flowering (commonly referred to as tasseling or arrowing in the sugarcane industry) for the varieties worked with in Hawaii at the time, field experiments were conducted to determine what quantitative effect tasseling might have on the yield of sucrose. The 1949 results of a replicated field test of ten paired plots using a current commercial variety supported the belief, long held by sugar growers, that flowering reduces sugar yields. In this field test, the average gain from suppression of flowering was 15 percent – the equivalent of 1.3 tons of sucrose per acre [133].

In the years immediately following these studies, the factors affecting flowering were extensively studied as were methods for preventing its occurrence. Effective ways found to prevent flowering included: night interruption with light, lower temperature, leaf and spindle trimming, withdrawal of water, and application of chemicals. Because temperature cannot be controlled in the field and because leaf trimming and light interruption on a commercial scale were not operationally feasible, emphasis was placed on water withdrawal and application of chemicals. Withdrawal of water was possible only on irrigated plantations and because such a practice had other operational problems, using chemicals eventually became standard practice in Hawaii.

The first potentially useful commercial chemical was maleic hydrazide, which gave about 60 percent control at best. Rapid developments led to the establishment of first monuron and then diuron as the chemicals of choice. Properly applied, 4 lbs/acre of either chemical gave virtually complete control of flowering in the heavy flowering varieties used in Hawaii during the 1950's and 1960's. Continued testing for active chemicals to prevent flowering led to the discovery that diquat applied from the air at rates as low as 0.125 lb/acre of the cation form was as active as monuron at 4 lbs/acre, making diquat one of the most active compounds yet evaluated for this purpose. The resultant cut in cost per unit acre for control was substantial. Tests included both the heavy-tasseling varieties in use at that time as well as the lighter-flowering varieties that were emerging as the preferred ones in the late 1960's. Positive effects with chemical control of flowering – with diquat being the compound of choice – have been obtained in Guyana, Mexico [134], the Philippines [135], and Taiwan [136], as well as in Hawaii.

Cotton is inherently a perennial plant that is cultured in practice as an annual crop. To obtain high yields, a continuation of growth and flowering and fruiting must be suppressed in order to facilitate defoliation and to aid in harvesting. As a result, growth regulators have been sought to induce an earlier termination of flowering and fruiting without adversely affecting yield. Studies with a number of regulators showed that chlormequat, the morphactins, and 4,4-dimethyl morpholinium chloride suppress late boll set but may also impair full development of the last bolls that matured. Conse-

Fig. 4-1. Inhibition of flowering in sugarcane by light interruption of night. Note absence of flowering around poles with light in an otherwise extremely heavily-flowering field

quently, the timing of application becomes a compromise between the desirable early termination of flowering and fruit set and the possible loss in yield and in quality of late opening bolls [137].

Early observations on the use of insecticides suggested that they might produce physiological effects on flowering and fruiting of cotton [138]. It has been reported that when cotton plants are grown to maturity under insect-free conditions, plants treated with certain insecticides such as guthion produce more flowers and more bolls and have a longer maturation period than do untreated plants [139]. On the other hand, some insecticides retard flower formation, boll set, and plant growth while hastening plant maturity [139]. Other studies showed that treatment with a toxaphene-DDT combination increased boll production each year but a concurrent reduction in boll size during the first year offset the boll number advantage and, therefore, yield was not affected. However, yield was increased the second year in the plots treated with toxaphene-DDT [140, 141]. There are conflicting reports from the effect of chlordimeform [142]. Work in Arkansas by Phillips and coworkers [143] showed beneficial effects on initiation and retention of cotton fruit; on the other hand, investigators in Mississippi found no difference in flowering rate, boll production, or yield. Conflicting results were also obtained in Mississippi from the use of acephate in different years, with an increase in flowering rate and yield obtained in 1977 but no effect in 1978 [138].

Fig. 4-2. Prevention of sugarcane flowering by airplane application of monuron (4 lb/acre). Fields in center and field in right foreground treated. Heavy flowering, untreated fields (white with flowers) in center background and left foreground. Waste land, far background

Termination of cotton flowering and fruiting by the use of growth regulators as a technique for controlling insects will be discussed in the chapter entitled "Resistance to, and Control of Pests".

Studies with growth retardants in treating woody perennial plants such as rhododendron and camellia showed not only a suppression of vegetative growth but a marked stimulation of flower bud formation [144, 145]. Later work with several types of holly showed phosphon to be more effective than chlormequat or dimethylamino maleamic acid [146].

Growth regulators affect the flowering of citrus. Studies with gibberellic acid applied during flower differentiation reduced the number of flowers per square unit of canopy but increased the number of leafy inflorescences and the percentage of fruit set [147–150]. On the other hand, daminozide application produced the opposite effect – an increased number of flowers per unit area of canopy and a reduction in the number of leafy inflorescences and percentage of fruit set [151, 152, 153]. These results with gibberellic acid confirm those of previous investigations. On the other hand, results with daminozide were mixed.

Spraying Montezuma roses in autumn with gibberellic acid caused an increase the following spring in the number of flowers and in the length, thickness, and fresh weight for the flower stems [154, 155]. One of the malformations occurring in roses is known as "bullhead" and is induced by suddenly falling temperature at early stages of floral development. This malformation is characterized by a decreased ratio of length-to-diameter of the floral bud, giving the flower a flat topped appearance, and is prevalent in the cultivar "Baccara". Injection of gibberellic acid into the receptacles of flowers grown at low temperatures prevents the symptoms of bullhead [156].

The application of gibberellic acid increased flowering in *Stevia rebaudiana*, the Paraguayan plant that produces stevioside, the sweetest naturally-occurring substance known [157]. Chlormequat delayed flowering in this plant [158].

Gibberellins are known to induce flowering in many temperate plants [159]. More recently, attention has been directed toward tropical plants. Gibberellic acid has been shown to induce flowering in several tropical aroids [160–164]. Work with woody monocots, showed both gibberellic acid

Fig. 4-3. "Earliness" and shorter plants in cotton through use of plant growth regulators. Left: treated; right: untreated. Photograph courtesy of U.S. Department of Agriculture

and the GA4/GA7 mixture to be effective in inducing flowering in *Cordyline terminalis* but ineffective with *Dracaena* spp. [165].

Promotion of flowering in conifers by the application of the gibberellins had been confined to the Cupressaceae and Taxodiaceae families until quite recently [166]. Such work primarily was carried out with gibberellic acid, apparently an unfortunate choice as this gibberellin is virtually inactive on members of the Pinaceae [167, 168]. This recent work has been conducted with a gibberellin A4/A7 mixture and found to be effective on numerous members of this family [168, 169], including several important forest tree species: Douglas fir [170, 171, 172], Sitka spruce [173–176], and *Pinus taeda* [177]. Sitka spruce is the most important economic forest tree in the United Kingdom, but it has poor natural regeneration because of the infrequency of good seed years; thus there is a need for a reliable method to increase seed production. Such an increase would also enable breeding programs to be carried out more efficiently.

Chapter 5. Gametocides

Much of the success in the United States with the production of corn is due to the successful commercial exploitation of hybrid vigor. About 90 percent of the North American acreage is now planted with hybrid varieties. Part of the commercial success is due to the discovery of male sterile cytoplasm which is used to prevent undesirable fertilization. As important as male sterile cytoplasm is, there is the constant danger that it might be incorporated into large areas of hybrid varieties allowing epidemic-like spread of plant diseases. This, in fact, is what happened recently in the United States where the use of the Texas male sterile cytoplasm led to the rapid spread of southern leaf blight through the commercial crop.

The use of a chemical gametocide would offer certain advantages over the use of cytoplasmic male sterility. Of particular importance among these advantages which have been listed by Batch [178] is (a) that male sterility could be produced as required, and (b) the avoidance of reliance on a single cytoplasm with its consequent dangers. These numerous benefits, some proved and some potential, have triggered research for gametocides.

The differential response of plants resulting in the formation of male or female flowers according to the regulator used, and the differential sensitivity to plant growth regulators of the male and female flowers already formed has led to the development of compounds aimed at the abortion of male structures. Naylor [179] was among the first to use maleic hydrazide to induce male sterility in corn plants; he suggested that this treatment might substitute for the hand-detasseling necessary for the production of hybrid corn seed. These results were confirmed by Moore [180]. Similar techniques were soon extended to cucurbits and tomatoes [181]. Seed production systems for hybrid grains depend on the use of male sterile cytoplasms. The use of such cytoplasms incurs danger from sensitivity to disease (as has been borne out recently in the corn belt of the United States). For some time it has been recognized that a chemical gametocide would offer significant advantages over the male sterile cytoplasm system.

A number of chemicals with demonstrated gametocidal activity have been evaluated. However, a chemical satisfying all the requirements necessary for a commercially usable gametocide have not yet been identified. Eight specifications have been listed by Batch [178] for the ideal chemical gametocide:

1) induction of male, but not female, sterility;
2) complete inhibition of pollen development;

Table 5–1. Plant Gametocides

Chemical Name	Common Name, Trade Name, or Designation	Company	Crop	Reference
1-(2-chlorobenzyl)-3-carboxy-4,6-dimethylpyrid-2-one	–	Rohm & Haas	cereals	183
2-chloroethylphosphonic acid	ethephon (Ethrel)	Union Carbide	wheat	182, 184–189
			barley	190, 191
			triticale	192
			sugarbeet	193, 194
			cucurbits	195, 196, 197
			millet	234
1-(p-chlorophenyl)-1,2-dihydro-4,6-dimethyl-2-oxonicotinic acid, sodium salt	RH-531	Rohm & Haas	barley	190, 200
			triticale	192, 199
			wheat	198, 199
3-(p-chlorophenyl)-6-methoxy-s-triazine-2,4-(1H,3H)-dione, triethanolamine salt	DPX 3778	DuPont	corn	201
			sugarbeet	202
			wheat	203
2-chloro-4-quinoline carboxylic acid	–	Union Carbide	corn	204
			tobacco	204
			wheat	204
2,4-dichloro-5-fluorophenoxyacetic acid	–	–	rye	205
2,3-dichloro-2-methyl propionic acid, sodium salt	Mendok (FW 450)	Rohm & Haas	alfalfa	206
			cotton	207, 208, 209
			rye	210
			ryegrass	211
			soybean	212
			sugarbeet	193, 213
			tomato	214, 215
			wheat	216

Table 5–1. (continued)

Chemical Name	Common Name, Trade Name, or Designation	Company	Crop	Reference
1,2-dihydro-3,6-pyridazine-dione	maleic hydrazide (MH)	Uniroyal	corn	179, 180
			grape	217
			pepper	218
			tomato	181
			wheat	216, 219
N-(m-fluorobenzyl)phthalimide	–	Monsanto	corn	220
4-fluoro-2,6-dichlorophenoxyacetic acid	–	–	rye	221
1-methyl-3-carboxy-4,6-dimethylpyrid-2-one	–	Rohm & Haas	cereals	222
Polychlorocarboxylic acid esters	–	–	wheat	223
2,4a,7-trihydroxy-1-methyl-8-methylene gibb-3-ene-1,10-carboxylic acid-1,4-lactone	gibberellic acid (GA_3)	Abbott	Brussels sprouts	224
		Elanco	cabbage	224
		ICI	cauliflower	224
			corn	225, 226, 227
			kale	224
			lettuce	228, 229
			sunflower	230
			wheat	216
2,3,5-triiodobenzoic acid	TIBA	IMC	grape	217
			tomato	181
			wheat	216
–	morphactins	Cela Merck	cucurbits	231, 232
–	RH-532	Rohm & Haas	wheat	233
–	UNI-D513	Uniroyal	triticale	192

3) independence from environmental conditions;
4) independence from genotypic differences;
5) wide flexibility of rate and stage of application;
6) absence of phytotoxicity or other adverse effects;
7) environmental safety; and
8) cost effectiveness.

Some of the better known gametocides are listed in Table 5–1 together with the crops on which they have been shown to be effective and the authors of the investigations.

In addition to the studies on crops, ethephon has been shown to induce male sterility consistently in a number of weeds: ragweed, redroot pigweed, lamb's quarters, barnyard grass, and crabgrass [182].

Chapter 6. Abscission

Controlling abscission (the separation or shedding of plant parts or organs such as leaves, flowers, fruits, or stems from the parent plant) is extremely important in agriculture and horticulture. To ensure the most effective crop growth, leaves usually should be retained in a healthy, green state. On the other hand, to simplify the mechanical harvesting of certain crops, such as cotton, it is highly desirable to have the leaves removed. A similar situation exists for fruit. In tree crops that have a large number of fruit started, it is sometimes desirable to thin the fruit by using an abscission-inducing compound, thereby increasing the size and quality of the remaining fruit. During crop growth, fruit should be retained on the tree for maximum development and maturity. However, at harvest for many crops the use of an abscission-inducing agent can be highly profitable by reducing labor requirements and cost; this is particularly true of citrus.

Many substances are known to affect abscission. However, few have been shown to have an effect on pectinase and cellulase in the abscission zone – the enzymes apparently required to dissolve the middle lamella thus causing the weakening of primary walls of cells in this zone that is necessary for abscission. Auxin, ethylene, and abscisic acid appear to be the hormones most directly involved, but through their effects on these three hormones, other compounds and other hormones can affect abscission indirectly. Apparently the control of abscission does not reside in any one hormone or environmental factor but is regulated by a complex interaction of environment, hormones, and physiological status of the plant.

The old practice of picking cotton – harvesting cotton bolls by hand – is well known but, generally, this practice has been replaced by machine harvesters. Since the introduction of the harvesting machine, the use of chemical agents as harvest aids has become commonplace. Harvest-aid chemicals of one type or another are used on more than three-fourths of the cotton acreage in the United States. Such chemicals are considered to be a necessity if cotton is stripper-harvested before frost and are considered to be advantageous if high quality fiber is spindle-picked from rank, heavily fruited plants with lush green foliage. The degree of success in obtaining economic benefits from the use of chemical harvest aids depends largely upon relationships between chemical, technique used, conditions of the crop, and climate. Chemicals now generally used as harvest aids on cotton may be classified as defoliants, desiccants, or other types of growth regulators [235]. Harvest-aid programs generally have showed definite economic advantages if established guidelines for the use of the chemical are followed, and such programs have become an important step in the production of high quality fiber [236].

After cotton defoliation was accidentally discovered in the late 1930's, research showed that it was feasible to remove leaves from a cotton crop before it was harvested. Interest in this procedure intensified with the advent of mechanical harvesters. Within a few years, thousands of chemicals had been screened in the search for efficient harvest aids. From this intensive search about twenty chemicals have been recommended for use. Of these twenty, only six are now extensively used. Two of the most widely used defoliants are the organophosphorous compounds, S,S,S-tributyl-phosphorotrithioate and tributylphosphorotrithioite. Defoliants are the preferred type of harvest aids to remove leaves without drying. Drying is undesirable because abscised leaves that are green and moist tend to fall free of the lint in the open bolls. Harvesting usually can be scheduled precisely after a defoliant is applied. Defoliants are used primarily where cotton is harvested by spindle-pickers, which cannot operate at peak efficiency if plants are rank and have an abundance of green, succulent foliage. Other defoliants in use are sodium chlorate, a highly flammable material when dry and usually formulated with a fire retardant, and manganese chlorate, a hygroscopic material thereby offering its own fire retardant properties. There is little difference in effectiveness between the chlorate defoliants: they work best on fully matured leaves and have little effect on young, immature leaves or on regrowth vegetation. Another commonly used defoliant is the arsenical cacodylic acid, formulated as the sodium salt. Its use is confined primarily to the western United States where cotton leaves are consistently tougher than those in the eastern states. Arsenic acid became an official candidate for Rebuttable Presumption Against Registration (RPAR) in 1978 [237]. Sodium chlorate with added endothal was selected by the Texas A&M Experiment Station as the defoliant combination to be used in their tests because sodium chlorate is the only cotton-harvest-aid chemical that is not a proposed candidate for RPAR by the U.S. Environmental Protection Agency [238]. A discussion of the toxicological reasons for governmental action against these compounds will be found in the chapter on toxicology.

Such regulatory problems make the search for new and effective compounds all the more important. A new defoliant, N-phenyl-N′-1,2,3-thiadiazol-5-yl-urea has been evaluated in field trials both in the United States and in South America [239]. In addition to its defoliant properties, this compound shows significant regrowth-inhibition following defoliation.

The discussion of desiccants as a substitute for defoliants as harvest aids for cotton will be found in the chapter on desiccation.

Materials of a formulation nature are added routinely to the harvest-aid chemicals to improve their effectiveness. These include the normal adjuvants: surfactants, spreaders, stickers, wetting agents, and the like. Recent research has shown the value of adding chemicals with either abscission- or desiccation-inducing properties to the defoliant mixture. Two compounds of this nature on which the most research has been conducted are endothal and paraquat [235].

A new experimental growth regulator, potassium 3,4-dichloro-isothiazole-5-carboxylate, has been evaluated extensively for its effectiveness in terminating late season vegetative and reproductive growth [235, 240, 241]. In these tests, it was found that this chemical increased the plant response to subsequent defoliant treatment. Results show that plants conditioned for increased response by treatment with this compound resulted in an increase of as much as 25 percent in defoliation without an adverse effect on yield and quality when this material was applied ten days before application of the defoliant.

Although abscission programs for citrus have been conducted in California, Israel, and Florida, the program in California has been greatly reduced because the market for California oranges is principally for high quality fresh fruit. Most chemicals that subsequently were found to affect the loosening of citrus fruit also caused some type of peel injury. On the other hand, over 90 percent of the Florida orange crop is utilized for processing; consequently, peel injury, unless it produces severe damage, is not important.

An active research program has been under way since the early 1960's in Florida, searching for compounds that are active in the abscission of citrus fruit. This program, initiated originally because of shortage of labor for hand picking citrus, is sponsored by the Florida State Citrus Commission working in conjunction with the Federal Government and with a number of industrial companies who supply synthetic chemicals for evaluation [242]. The first compound found to produce abscission of citrus fruit was iodoacetic acid, which caused severe phytotoxicity problems [243, 244]. Many of the early active compounds, including certain weak acids, were of the "mass action" type [245], so called because they were applied at high rates (up to 100 pounds per acre); rates such as these damaged both the fruit and the trees. A major breakthrough occurred when an antifungal antibiotic, cycloheximide, was found to loosen citrus fruit at application rates of less than 0.1 lb per acre [246, 247]. Although cycloheximide is used commercially in Florida on most orange varieties, it is not used during the harvest of Valencia oranges because it damages the flower and the immature fruit [248, 249]. Efforts to find suitable abscission chemicals for Valencia oranges faced a number of problems but development of a selective abscission material, 5-chloro-3-methyl-4-nitro-1H-pyrazole (Release), provided effective induction of abscission in mature Valencia oranges without damaging either new twig growth or immature fruit [250–254].

Because of increasing shortages and increasing costs of labor in Australia, a similar situation exists in that citrus industry regarding the need for mechanical harvesting of citrus fruit and the desire to reduce the relatively high attachment force of citrus fruits. The studies in Australia concentrated on three materials: ethephon, cycloheximide, and 5-chloro-3-methyl-4-nitro-1H-pyrazole. Although ethephon promoted abscission without visible injury to the fruit, the excessive leaf drop that accompanies the treatment limits its desirability. The results with cycloheximide showed leaf and fruitlet drop in addition to that for mature fruit; this effect was too high to warrant its use in commercial practice. "Release" appears to be selective for mature fruit only and was found to be the compound most promising for incorporation into harvesting systems [255].

More recent work in Florida has been with combinations of the more active abscission compounds [256, 257]. In fact, the best combination was a three-way one with Release + cycloheximide + chlorothalonil (Sweep) [258]. Not only did this combination give better abscission of the fruit than any other combination, but required 25 to 50 percent less total chemical compared to individual chemical usage – thus showing synergism.

Because of the importance of the Valencia orange, which constitutes the major part of the Florida crop, studies continue in an effort to find effective abscission agents. One of the more promising compounds recently found to be effective for Valencias is ethanedial dioxime (glyoxime or Pik-Off) [259, 260].

Manpower availability for olive picking is becoming an increasing problem throughout the world for olive growers. Considerable research is being carried out in the Mediterranean region [261–264] and in California [265, 266], the main goal being to decrease the amount of labor needed for olive harvesting. Efforts to decrease labor consumption are directed toward a mechanical solution, a chemical solution, or a combination. A number of compounds have been found to promote olive abscission but only under certain ambient conditions [265, 267–270]. Of the materials tested, ethephon has proven to be one of the most effective chemicals causing the abscission of olives without promoting excessive leaf drop [262, 271–274]. The most recent effective compound to be added to this list is 2-chloroethyl-tris (2-methoxyethoxy) silane (Alsol) [263, 266, 275], found under certain circumstances to be as effective as ethephon. Consequently, olive growers now can choose from two chemicals of similar efficiency. Obviously, the decision will be made based on economics, availability, and local effectiveness.

Olive varieties in California bear fruit in cycles of heavy yields followed by light yields. Sometimes in the light year the production is so depressed that harvest is not economical. A solution to this problem has existed for many years – the use of naphthalene acetic acid. However, California growers have been reluctant to thin with NAA, probably because of difficulties in determining fruit size for the timing of treatment and the possibility of high temperature following treatment. The recent work in California has shown several methods of better estimating the timing of application for maximum effectiveness [276].

Removing excess fruits from apple trees is an essential orchard practice [277]. This practice has two favorable results: 1) it reduces biennial bearing and, 2) it increases the fruit size, color, and quality of the fruit taken to harvest. If the treated trees have sufficient vigor, then fruit size at harvest is directly related to the earliness and the degree of fruit thinning. Prior to 1940, apples were thinned by hand. The first attempt at chemical elimination of flowers in apples was made in 1934 using common spray materials available at that time such as copper sulfate, zinc sulfate, calcium polysulfide, sodium

Table 6–1. Chemicals Used for Apple Thinning

Chemical	Common Name or Designation	Trade Name(s)	Other Use(s)
2-chloroethylphosphonic acid	ethephon	Ethrel Cepha	plant growth regulator
Methyl N′,N′-dimethyl-N-[(methyl carbamoyl) oxy]-1-thiooxamimidate	oxamyl	Vydate	insecticide, nematicide, acaricide
Naphthaleneacetamide	NAD NAAm	Amide-Thin Anna-Amide	plant growth regulator
1-Naphthaleneacetic acid	NAA	Fruitone Fruit Fix Fruit Set Plucker Stafast Kling-Tite Tip Off Tre-Hold	plant growth regulator
1-Naphthyl-N-methylcarbamate	carbaryl	Sevin Dicarbam	insecticide
Sodium 4,6-dinitro-ortho-cresylate	DNOC DNC	Elgetol Chemsect Sinox	insecticide, fungicide, herbicide
β-chloroethylmethyl-bis-benzyloxysilane	CGA 15281	–	plant growth regulator

polysulfide, and oil emulsions [278]. By 1940, two materials showed promise in reducing fruit set in apples: dinitrocyclohexophenol and sodium 4,6-dinitro-orthocresylate (DNOC) [279, 280]. There followed several years of adaptation of DNOC as a bloom spray for thinning apples. During the 1940's and 1950's studies were conducted with NAA and naphthaleneacetamide (NAAm) [281, 282]. Studies were conducted to obtain both the proper concentration to be used and the proper timing. By the mid 1940's, it had been shown that NAA could be used as a post-bloom spray [283]. Interest in this compound waxed and waned until the mid 1950's when interest was revived and more definitive practical answers were obtained. As a result, DNOC and NAA or NAAm came into commercial use on apples in many areas of the United States.

Table 6–2. Effect of Combination Sprays of Chemical Thinners with Plant Growth Regulators on "Golden Delicious" Apple Trees (after Williams, 277)

Treatment	Fruit per Blossom Cluster Number	Subsequent Years Return Bloom Percentage
Control	95	2
DNOC	67	4
DNOC + NAAm + carbaryl	56	23
NAAm + carbaryl	75	16
DNOC + NAAm + daminozide	59	26
NAAm + daminozide	71	28
DNOC + NAAm + ethephon	43	72
NAAm + ethephon	66	35

In the late 1950's, a then-new insecticide, 1-naphthyl-N-methyl carbamate, commonly known as carbaryl (Sevin) reduced fruit set when applied after full bloom for control of insects [284–287]. Naturally this led to a study of the fruit-thinning properties of carbaryl on apples, and it was found to be a highly effective agent [288–293], thus adding another chemical available both for research purposes and for commercial use. Since that time, the only chemical added to the registered list for thinning apples is ethephon, used primarily on "Golden Delicious" apples [294]. Studies in Australia showed ethephon to be superior to carbaryl or NAA for thinning of "Jonathan" apples with no disorders in the treated fruit [295]. Those chemicals most commonly used in the United States for apple thinning are given in Table 6–1.

The latest additions to the group of apple thinners are another insecticide, oxamyl (Vydate) [290], and a new ethylene-releasing compound, CGA 15281 (β-chloroethylmethyl-bis-benzyloxy silane) [289]. More recent work dealing with thinning of apples is concerned with variations in cultivar response, with differences in climatic areas, and with combinations of the growth regulators.

The problem of biennial bearing has been successfully overcome by combining chemical thinning agents such as NAAm with other growth regulators such as ethephon or daminozide [277]. They can be combined as a tank-mix and applied as a post-bloom thinning spray. The more promising combinations are shown in Table 6–2. The choice of treatment depends upon the amount of bloom and the thinning needed.

Although the selective removal of both blossoms and small fruits in apples is routinely accomplished with plant growth regulators in practically all producing regions, it is probably the most highly developed in western North America. The early thinning increases the size of the remaining fruit and improves other aspects of fruit quality in addition to providing a better fruit appearance. Another, and often more important, reason for thinning of flowers and fruit relates to the biennial bearing tendency of many apple cultivars. The profound influence of chemical thinning agents on alleviating biennial bearing of apples can be seen by examining national production records before and after thinning practices came into widespread use. Before 1949, biennial bearing was a national problem in the United States. Since that year, it has largely disappeared.

Results with peaches have not been nearly as satisfactory as those with apples. The search for an effective agent for thinning peaches has not yet resulted in a commercially accepted method of fruit removal [296, 297]. Active, but not yet acceptable, compounds include 3-chlorophenoxy-α-propionamide [298–302], naphthaleneacetic acid, N-1-naphthylphthalamic acid, β-chloroethylmethyl-bis-benzyloxysilane [296, 303–306] ethephon [300, 302, 307–312], 1,1,5,5-tetramethyl-3-dimethylaminodithiobiuret [308, 313], 3-chloroisopropyl-N-phenyl carbamate [314, 315]. In addition to its phytotoxic side effects on peaches as well as other stone fruits, ethephon both over- and underthins peaches [294, 302, 316]. The simultaneous application of gibberellic acid with ethephon either eliminates or significantly reduces the undesirable side effects of ethephon without altering thinning response [317]. The use of β-chloroethylmethyl-bis-benzyloxysilane and an analog of the same compound caused thinning responses that varied with cultivar, timing, concentration, and year [296, 318]. In addition, unpredictable and commercially unacceptable leaf abscission occurred with most cultivars throughout the test period.

Ethephon appears to do a more satisfactory job in the thinning of plums [319–323] and cherries [321–324] than it does for peaches. It is also used effectively as a loosening agent in sweet cherries to aid in mechanical harvesting [325]. NAA and 2,4,5-T also have been found to be effective in thinning of plums [326]. Chemical thinning of oranges has been successfully carried out with NAA and ethephon (Valencia) [327, 328], and with NAA [329, 330] and ethyl-5-chloro-1H-indazol-3-acetic acid ethyl ester (Mandarin) [331].

Ethephon has been used effectively in the spraying of Macadamia trees to induce nut drop in Australia [332]. In California, the same chemical has been effective in permitting an earlier harvest of walnuts [333].

a

b

Fig. 6-1 a, b. Effect of defoliant on cotton plants. **a** Well defoliated cotton. **b** undefoliated cotton. Photographs courtesy of U.S. Department of Agriculture

For grapes under conditions conducive to fungal infection, particularly in compact clustered varieties, the use of growth regulators as thinning agents is advisable [334]. Commercially, the only growth regulator used at present in California is gibberellic acid applied prior to bloom. Even though other chemicals such as naphthaleneacetic acid have shown promise [335–338], they are not used commercially [334]. Recent work in California is aimed at finding the appropriate time and concentration for the application of naphthaleneacetic acid on "Carignane" grapes [334]. In this work they find there is a period between bloom and several weeks after bloom when vines can be sprayed with naphthaleneacetic acid without danger of

Fig. 6-2. Mature "Valencia" orange fruit drop following early june application of 5-Chloro-3-Methyl-4-Nitro-1 H-Pyrazole. Note leaf and immature fruit drop are negligible. Photograph courtesy of Florida Department of Citrus

reducing crop yield so much that it would be undesirable. The optimal concentration varies according to the developmental stage of the treated vines.

Ethephon has been used to aid in the abscission of grapes [339–342]. The principal problem with this chemical is in determining a concentration that would give the desired loosening effect without dropping berries on the ground before harvest and without adversely affecting the leaves. The optimal concentration varies with the cultivar, environmental conditions, and surfactant used [343].

Overbearing in the early stages is a problem in young coffee trees. This is often the result of excessive flowering followed by abundant fruit set which can tax the overall energy of young coffee plants – sometimes to the extent that later productive capacity is impaired. Ethephon has been found to produce large reductions in the percent of fruit retained [344, 345].

Japanese persimmon fruit can be thinned with naphthaleneacetic acid if applied within 30 days after full bloom. Supplementary treatment with gibberellic acid almost completely negates this thinning action [346, 347, 348].

Chapter 7. Fruit Set and Development

For many agricultural crops, the ability to control the amount of fruit set is of great importance. For those crops whose yield is limited by their ability to set fruit, increasing the incidence of fruit setting is the key to increased yields. For those crops where an over-abundance of fruit has been set, the need is to carry out effective, safe thinning operations. The use of plant growth regulators in thinning operations is discussed in the chapter on abscission. There has been an increasing use of plant growth regulators to control fruit set as well as to influence the subsequent development of fruit – including size, shape, and rate of maturation. Compounds most frequently used for the induction of fruit set are of the auxin type.

Fruit size is generally connected with marketability. Usually, larger fruits are preferred by the consumer. There are instances, however, when a decrease in size is desirable. The timing of the maturity cycle is often important in marketing the product as well as the ability of the grower to remove the crop during a given time period. This approach is discussed with regard to certain crops, such as oranges, in the chapter on metabolism.

Some plants (such as the cultivated banana, navel orange, oriental persimmon, pineapple, and certain varieties of fig and pear) develop fruits in nature in the absence of pollen. Other plants, such as certain grape varieties, develop fruit from the stimulus provided by the presence of pollen alone without fertilization of the ovule, presumably because of the secretion of hormonal substances.

Scientists have found that many synthetic growth regulators aid in developing fruit in plants. The best of these are 4-chlorophenoxyacetic acid and 2-naphthoxyacetic acid. These chemicals are most effective on fruits that have many ovules – such as tomato, squash, eggplant, and fig. However, these chemicals are usually ineffective on peach, cherry, plum, and other stone fruits. Many fruits that can be set by such hormonal compounds also can be set by the gibberellins. In addition, gibberellins can set fruit in some species that do not respond to the other chemicals. An extensive study concerning the use of growth regulators to improve fruit set of citrus showed gibberellic acid to be the growth regulator holding the most promise [349]. Other work has shown that aqueous sprays of gibberellic acid to entire trees in full bloom increased both set and yield of seedless fruit of five self-incompatible citrus cultivars. Seedless fruit were, however, more likely to drop after initial fruit set than seedy fruits that result from cross-pollination [350].

Girdling has long been used to increase fruit set *in grapes* [351, 352]; to this practice the added benefits of using plant growth regulators (especially growth inhibitors) have been demonstrated. Generally, these materials are applied just prior to full bloom to improve berry set in grapes [353]. However, cultivars of grapes differ in their response to the various growth retardants [354]. For example, daminozide was found to be most effective on "lambrusca" cultivars [355], whereas chlormequat was more useful with "vinifera" cultivars [356]. Chlormequat significantly increased the set of "Himrod" cultivars as well as "de Chaunac", although there was a deleterious effect on berry quality [354]. Gibberellic acid applied as a post-bloom spray increased the set of "Himrod" grapes and reduced their acidity [354]; at full bloom, set was not improved [357]. Results from Iran have shown daminozide to increase the number of berries per cluster as well as the cluster weight [358]. This work, as well as previously reported studies, shows that the best fruit set in grapes is obtained by a combination of girdling and the use of the proper growth regulator [358]. Kinins also have been shown to be effective in fruit set of grapes [359] and, in some instances, to have a positive effect on subsequent fruit growth [360].

Yield improvement for the *cucumber plant* is directly related to increased fruit set on individual plants. Generally, cucumbers do not set more than two fruits per plant at any one time because the earliest ones set restrict the development of additional fruits [361]. Certain growth regulators, including chlorflurenol, effectively overcome this limitation in fruit set [362, 363, 364]. However, for chlorflurenol to be totally effective the plant must have predominantly female flowers. Ethephon induces female flowering in the cucumber and thus should enchance the potential for chlorflurenol to enhance yield [365], a supposition now confirmed [366]. In cucumbers previously treated with ethephon, chlorflurenol increased the number of fruit in the smaller size groups in all five pickling cultivars tested, thus promoting an increase in dollars per acre. It also reduced the percentage of off-shaped fruit, again increasing the value in dollars per acre [366]. Recent studies on the regulation of fruit set in cucumber suggest that the use of plant growth regulators should provide the potential for every viable pistillate flower at the time of application to set and develop.

The *eggplant* is a warm season vegetable. Attempts to extend its production period during winter and spring, even under plastic covers, were unsuccessful until the plant growth regulators such as β-naphthoxyacetic acid and n-meta-tolyl-phthalamic acid were applied to assist both in fruit set and in fruit development [367].

The yield of *cashew* is quite low, largely because of poor fruit set and high rate of fall of immature nuts following fruit set. Studies in India with many commercial plant growth regulators showed all of those tested to increase fruit set – with 2,4-D, NAA, and IBA being the best [368].

Probably the three most important factors affecting yield in *aestivum wheats* are tillering, number of grains set per ear, and grain size attained. Studies with growth regulators in India showed that gibberellic acid applied during grain setting led to an increase in grain yield because of an increase in the number of grains per ear, whereas applying indoleacetic acid increased the yields because of the increase in size of the grain [369].

Anjou pears are known for taking a long time to reach maturity for bearing. Usually trees produce little fruit until the sixth or seventh year after planting and take ten to eleven years to reach full production. Work by U.S. Department of Agriculture investigators with plant growth regulators suggests that, if treated properly with the appropriate growth regulator, Anjou trees can be made to bear fruit two to three years sooner and with higher production. These investigators found that chlormequat increased fruit set after one application; after two consecutive years of applications, yield was increased about 50 percent [370]. The two other growth regulators studied in these tests, ethephon and daminozide, required two seasons to be effective. Treatment with aminoethoxyvinylglycine effectively retains the flesh firmness of apples [371] and pears [372]. A more recent report confirms this effect on firmness in apples and also shows the complete elimination of the "June drop", thus significantly increasing fruit set for the following year [373]. Studies with gibberellic acid on pears in Hungary showed a generally favorable effect on fruit set and market quality with applications of 10 to 50 parts per million. These favorable effects were observed only in years in which pollination conditions were unfavorable or when flowers were frost-damaged [374].

Studies with *tomatoes* in Mexico showed increase of fruit set in greenhouse-grown tomatoes by applying phenoxyacetic acid at 70 parts per million [375]. Work with tomatoes in Poland show that morphactant applications at the proper dose level can cause an increase in both the total number and the weight of fruits in the "New Yorker" variety [376].

2-Amino-6-methylbenzoic acid increased the fruit set of both *"Golden Delicious" and "Jonathan" apples*. It also reduced June and pre-harvest drop [377]. Applying daminozide in June, in combination with pruning, increased fruit set in the second season for "Jersey Mack" apples [378].

Being able to control the shape of fruit has practical implications. For example, apple cultivars known to have superior internal quality attract premium prices if the market can identify the product by some characteristic feature of its external appearance. The flavor, aroma, and quality of the "Red Delicious" apple is world-renowned. In the marketplace, this cultivar is identified primarily by the elongated appearance of the fruit and the presence of pronounced calix lobes ("crown"). Fruit having these attributes is said to be "typey". Commercial application of a growth regulator mixture that is a combination of gibberellins A4 and A7 with a cytokinin 6-benzyladenine, named Promalin, caused improved "typiness" in "Red Delicious" apples [379, 380, 381]. The timing of application is the most important factor in producing these desired results. The period between full bloom and early petal fall is the time when flower parts are most receptive to the treatment and are capable of absorbing and transferring these regulators to the developing fruit. The success with apples has resulted in this combination product being tested for a number of different uses in a wide variety of crops.

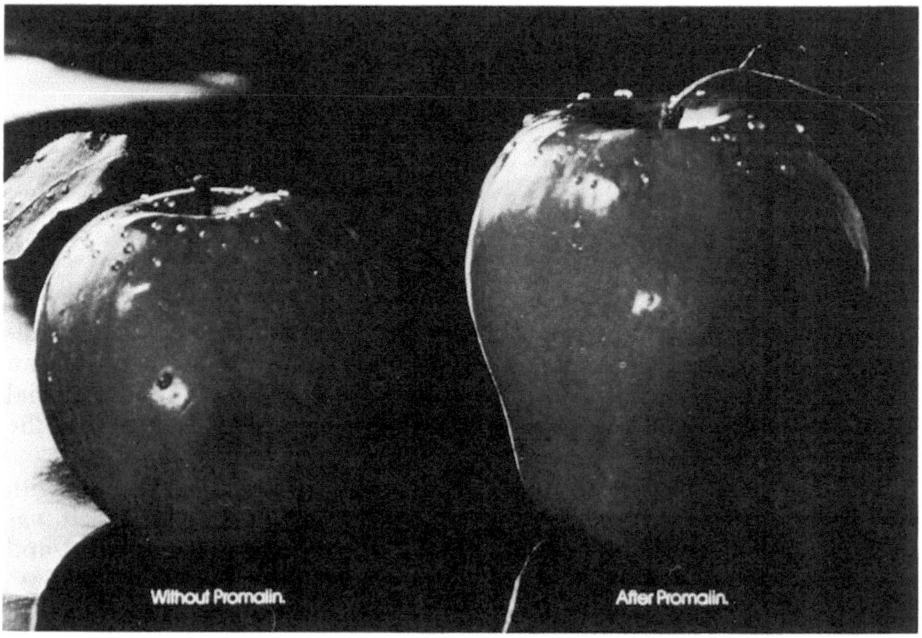

Fig. 7-1. Effect of combination gibberellins A 4 plus A 7 with a cytokinin (6-Benzyladenine) on apple fruit shape and size. Left, untreated; right, treated. Photograph courtesy of Abbott Laboratories

Fig. 7-2. Effect of ethephon in promoting boll opening in cotton. Treated part of field in foreground; untreated portion in background. Photograph courtesy of Union Carbide Agricultural Products Company

Chapter 8. Plant and Organ Size

The control of plant and organ size can be of great importance in agriculture. If maximum weight, length, or diameter affects final yield, then an increase in size is desirable. On the other hand, if it can be of commercial benefit, it may be important to be able to reduce the overall size of the plant.

The elongated "foolish seedling" effect in rice, caused by infection with the fungus *Gibberella fujikuroa*, has been known for many decades. It was not until 1938, however, that a metabolite of this fungus was isolated and shown to be the causative agent of the disease. The isolation, crystallization, and structural determination of this material by Yabuta and Sumiki in 1938 [382] led to the discovery of a new class of hormones, the gibberellins. Because of wartime secrecy, little was published about these new "wonder compounds" until World War II had ended. Then, in the 1950's, publications both from Japan and the United States reported the spectacular effects on both ornamental and edible crop plants.

In most plants, the outstanding effect of the gibberellins is to elongate the primary stalk. This effect occurs in the young tissues and growth centers and is caused by an increase in cell length, an increase in the rate of cell division, or a combination of both, depending on the specific types of plant treated. Gibberellins have remarkable effects on many dwarf plants including peas, corn, and beans; when treated with gibberellins, these plants grow to full size. Also, gibberellins (a) affect the extent to which the plant develops side branches, and (b) increase the size of many young fruits, especially grapes. Because gibberellins induce the production of the enzyme amylase in barley, they are commonly used in the malting of this grain.

Although the gibberellins can induce flowering in many plant species, their greatest commercial uses have been in increasing the size of grapes and in stimulating the growth of sugarcane to increase the length of the primary stalk. Treatment of sugarcane with as little as 2 ounces of gibberellin per acre increases the yield of cane more than 5 tons per acre and raises the output of sugar as much as 0.5 ton per acre [131, 383, 384]. Reports from Hawaii show that maximum increases in both length and fresh weight of seven commercial Hawaiian cultivars were a function of a cultivar treated, the amount of gibberellin applied, the number of applications, and the interval between applications [385, 386]. The gibberellins have been tested also on a wide range of vegetable crops with results so far that are spectacular but often not beneficial. For example, in many species the treatment can induce premature flowering, which is undesirable in crops such as cabbage because it is the vegetative leaves that are commercially valuable.

In 1977, investigators at Michigan State University reported spectacular results for a number of crop plants whose growth can be greatly stimulated by a naturally occurring long-chained alcohol named triacontanol [387, 388], which they isolated from alfalfa. Because this alcohol increased the dry weight of test plants grown in the dark, it cannot have an effect on photosynthesis. These and other investigators speculated that triacontanol might increase the uptake of nutrients [389–394]. Since those results were published there have been numerous reports in the literature concerning triacontanol, many of them failing to show effects in the test systems used or on the crops studied [394–401]. Whether or not triacontanol will be of real value to agriculture cannot be decided at this time. Because of these inconsistent results the original investigators are attempting to determine the factors that might be responsible for the lack of response reported in other investigations. The original investigators found that other longchained primary alcohols [402] and other chemicals, such as morpholine, inhibit the activity of triacontanol; they found that acidity adversely affects triacontanol activity – to show activity triacontanol must be administered in solutions above pH 7 [403].

Japanese investigators have found that 4-ethoxy-1(*p*-tolyl)-*s*-triazine-2,6) (1H:3H) dione promoted mesocotyl growth in dark-grown rice seedlings [404, 405]. With cultivars of the *japonica* type, this compound by itself showed a small promotive effect; when combined with gibberellic acid, it had a marked synergistic effect on growth. In cultivars of the *indica* type, the compound alone showed considerable promotive effect, but when combined with gibberellic acid only an additive effect. Earlier work had shown that this compound is a powerful synergist of gibberellic acid in promoting shoot growth of rice seedlings grown in the light [406].

The first steroid shown to accelerate plant growth was purified and identified by USDA scientists [407]. Named *brassinolide*, the steroid was isolated from the pollen of the rape plant [408, 409]. Structural isomers of the naturally occurring steroid have been synthesized, and several of these have been found to be active [410].

On the other hand, there are circumstances where reducing the size of the overall plant could be of commercial benefit. An example is the reduction of stem lengths in cereal crops by *chlormequat*; use of this compound has become an important factor in farming, especially in the growing of wheat in Europe [411]. Chlormequat prevents, or greatly reduces, the probability that wheat will "lodge" (fall over) in heavy winds and rain. Most of the wheat grown in West Germany is treated with this compound; in fact, that country's increased yield of wheat is largely attributable to use of this plant growth regulator. Because of this particular use, chlormequat is among the most widely used plant growth regulators in the world [412]. Sometimes when no lodging occurs, chlormequat still increases the yield of wheat; this is accomplished by increasing the number of ears and the number of grains per ear, probably because of an increase in shoot number

[413]. Daminozide acts as a general inhibitor of growth in peas, lowering the number of the node in which flowers are formed [414]. Foliar application has little effect on vegetative characteristics of processing pea seedlings but increases the yield of "Early Perfection"; the compound has no significant effect on three other cultivars.

By reducing the amount of lodging, the combination of mepiquat and ethephon shortens the height, improves the standing power, and increases the yields in winter wheat [415, 416].

Mowing *turf grass* is a time-consuming and costly maintenance procedure for both homeowners and professional turf managers. Of considerable interest is the feasibility of using chemical growth retardants to reduce the mowing frequency and management costs of such turf areas as highways, rights-of-way, parks, cemeteries, golf courses, and lawns. Maleic hydrazide has been used since the 1950's for cool season regulation of grass growth. Chlorflurenol, fluoridamid, mefluidide, ethephon, and ancymidol all have been evaluated for their effectiveness – alone and in combination – in suppressing both vegetative growth and seed-head formation of grasses [417–428]. Probably the most studied compound of this group is mefluidide, which is effective under a number of conditions; however, at effective rates, there are many reports of phytotoxicity [429]. The use of turf growth retardants continues to be commercially attractive and several companies have continuing programs to evaluate experimental compounds in the field (see Table 8–1).

The development of growth retardants has had a considerable impact on the production of *floral crops* such as chrysanthemum, poinsettia, and other decorative flowers [450]. The most commonly used growth retardants for this purpose are daminozide [451] and chlorphonium chloride [452]. A new quaternary ammonium retardant has become available for research purposes and commercial formulations of this material, 1-allyl-1(3,7-dimethyloctyl)piperidinium bromide (piproctanylium bromide) [452, 453, 454], have been granted clearance for commercial trials. Extensive investigations have been carried out with ancymidol, particularly for poinsettias [455, 456], and with dikegulac sodium for preventing shoot elongation in junipers [457].

Cotton is one of the most important commercial crops. Regardless of the amount of available water, because it is a perennial the cotton plant tends to produce more leaf, stem, and flower material than is necessary for a given fiber yield. This tendency has a restrictive influence on the planting of a greater crop density, on the optimal use of fertilizer, and on the introduction of more efficient harvesting techniques. Only in regions with irrigation is the grower able to partially reduce cotton's undesirable rank growth by regulating the amount and timing of the water supply. Investigations have shown plant growth regulators to be of help in solving such problems. One useful chemical is 1,1-dimethyl-piperidinium chloride (mepiquat chloride), which causes a cotton stand to develop a healthy dark green color within a few days after treatment [459–462]. Inhibiting cell elongation and node formation results in decreased growth of the cotton plant in both length and

Table 8–1. Grass Retardants

Chemical Name	Designation	Common Name	Trade Name	Source	Reference
1,2-Dihydro-3,6-pyridazinedione	MH	maleic hydrazide	Slo-Gro	Uniroyal	417–422, 424–426, 430–434
2,3-Dihydro-5,6-diphenyl-1,4-oxathiin	UNI-P293	–	–	Uniroyal	417, 432, 434, 435
2,3-Dihydro-5-(4-methylphenyl)-6-phenyl-1,4-oxathiin 4-oxide	–	–	–	Uniroyal	435
3-Trifluromethyl-sulfonamido-p-acetotoluidide	MBR-6033	fluoridamid	Sustar	3M	417–422, 424, 426, 432, 434
N-2,4-dimethyl-5-(trifluromethyl)sulfonylamino-phenyl acetamide	MBR-12325	mefluidide	Embark	3M	417, 422–424, 426, 429, 432, 434, 436–439, 458
Not released	CGA-17020	–	–	Ciba-Geigy	417, 432, 434, 440
2-Chloro-N-(2-ethyl-6-methylphenyl)-N-(2-methoxy-1-methyl-ethyl)acetamide	CGA-24705	–	–	Ciba-Geigy	417, 432, 434, 440
N-phosphonomethyliminodiacetic acid	MON-820	–	–	Monsanto	417–421, 432–434, 441
2-Amino-6-methyl benzoic acid	ACR-1308	–	–	Maag	442, 443
Not released	PP-333	–	–	ICI	–
Methyl 2-chloro-9-hydroxyfluorene-9-carboxylate	IT-3299	chlorflurenol	(Morphactin)	E. Merck AG	418, 419, 421, 424, 425, 431, 444
Methyl 9-hydroxy-fluorene-9-carboxylate	IT-3235	flurenol	(Morphactin)	E. Merck AG	418, 419, 421, 425, 444
Methyl 2,7-dichloro-9-hydroxy-fluorene-9-carboxylate	IT-3353	dichlorflurenol-methyl	(Morphactin)	E. Merck AG	418, 419, 421, 425, 444
7-Oxabicyclo (2,2,1)heptane-2,3-dicarboxylic acid	–	endothal	–	Pennwalt	425, 445–449
2,3:4,6-bis-O-(1-methylethylidene)-O-(L)-xylo-2-hexulofuranosonic acid, sodium salt	–	dikegulac sodium	Atrinal	Maag	423
N,N-bis-(phosphonomethyl) glycine	MON-845	glyphosine	Polaris	Monsanto	418, 419, 421, 433, 441

Table 8–1. (continued)

Chemical Name	Designation	Common Name	Trade Name	Source	Reference
Ethanolamine-p-nitrobenzenesulfonylurea	MON-814	–	–	Monsanto	418
Phosphonomethylglycine	MON-139	glyphosate	Round-Up	Monsanto	418, 419
Phosphonomethylglycine, calcium salt	MON-464	–	–	Monsanto	418, 419
Not released	C-19490	–	–	Ciba-Geigy	419
2-(1-[2,5-dimethylphenyl]-ethylsulfonyl)-pyridine-N-oxide	–	–	–	Uniroyal	448
2-(1-phenylethylsulfonyl) pyridine-N-oxide	–	–	–	Uniroyal	448
2-Amino-4-methoxy-3-butenoic acid	–	–	–	Ciba-Geigy	449
α-Cyclopropyl-α(p-methoxyphenyl)-5-pyrimidine methanol	EL-531	ancymidol	A-Rest	Elanco (Lilly)	417

breadth. This inhibition of growth is particularly marked in the upper plant parts that are not as important for the production of yield. Cotton plants so treated have a more compact conical form, allowing the plant to be spaced closer with narrower rows and a higher plant population. The end result appears to be a boll set with considerably higher weight, giving greater yield at the first picking.

Extension growth in *grapevines* is usually quite strong and demands repeated topping. Because it stimulates growth of lateral buds, topping is a continuous operation [463, 464]. Results of studies with chlormequat and daminozide were unsatisfactory [465, 466]: extension growth was not restricted enough and lateral buds were only partially inhibited; in some cases fruit growth was also inhibited. Studies in Israel [467] showed ethephon at 480 ppm efficiently inhibited the terminal growth of cane in five grapevine cultivars and prevented the opening of lateral buds on the shoots for about 8 to 10 weeks. No negative effect was found in the following season. The results with ethephon were considerably better than the other compounds evaluated such as morphactins, daminozide, and others. The results suggest ethephon treatment to be more efficient and uniform than the repeated manual topping.

The yield of *raspberry plants* is influenced by a number of factors including size of cane, cane growth, and cane removal or suppression [468, 469, 470]. The differences in productivity of canes is believed to result from varying availability of nutrients, or competition for the nutrients while blossoms and berries are developing during the spring. Suppression of primocanes by spraying with 4,6-dinitro-sec-butylphenol (dinoseb), after the canes had reached about 18 cm in height, for four successive years resulted in an increased yield each year [471].

The history of growth retarding chemicals is longer than that of stimulants, and the success is greater. Cathey [472] earlier defined plant growth retardants as synthetic organic chemicals that, when applied to responsive plants, reduced the rate of stem elongation, usually without producing substantial growth malformations. This definition remains broadly accepted, allowing a distinction to be made between this constantly growing group of compounds and other types of plant growth inhibitors. The capacity of retardants to restrict stem growth without reducing leaf number, and usually without substantially reducing leaf area, is reflected in the compact growth habit of treated plants and is the main reason for the commercial success of retardants in floriculture and in agriculture.

The effect of retardants on stem growth has been traced to the subapical region of the shoot tip where cell division and, to a lesser extent, cell elongation are inhibited. Thus, internodes of retardant-treated plants are shorter primarily because they posses fewer cells. Many growth retardants have been shown to act by inhibiting a specific step in the synthesis of naturally occurring gibberellins which are necessary for the maintenance of subapical meristem activity. When such retardants are used, it is possible to reverse the inhibition effect in intact plants by the simultaneous, or later, application of the appropriate dose of gibberellic acid.

Maleic hydrazide may be distinguished from the growth retardants in that it inhibits cell division both in the apical and in the subapical meristem; it thus not only suppresses stem growth but prevents the initiation of leaves. Consequently, it is unsuitable for control of shoot growth in floriculture, but has been used extensively for suppressing unwanted vegetative growth in grasses and for controlling sucker growth in tobacco. It is considered to be both an anti-auxin and a gibberellin antagonist.

The morphactins produce a wide range of morphological effects including suppression of apical growth usually accompanied by promotion of lateral bud development [444]. Such effects may be explained because morphactins are potent inhibitors of polar auxin transport upon which apical dominance depends.

Ancymidol has been found to be effective for a wide range of *woody and herbaceous plants* including both temperate and tropical cultivars. Work in California showed it to have a wide latitude of safety on most plants [473, 474]. When objectionable effects were observed, these were thought to be due primarily to the result of high dosages used in the experimentation. In Florida, investigators looking for growth regulators to reduce the internode elongation in tropical foliage plants found ancymidol to have the widest spectrum of activity at the level studied [475, 476]; both soil and foliar applications were found to be effective.

Fig. 8-1. Effect of gibberellic acid on growth of bean plant. Left, untreated plant; right, gibberellic acid-treated plant

GA
X

Fig. 8-3. Effect of gibberellic acid on grapes. Left, untreated; right, treated. Photograph courtesy of Abbott Laboratories

◄ **Fig. 8-2.** Effect of gibberellic acid on growth of sugarcane. Stalk on left treated with 4 oz/acre of gibberellic acid; stalk on right untreated. Note three greatly extended nodes in treated stalk, both stalks are being held by the same node, which is the last fully developed node at time of treatment

Fig. 8-4. Effect of gibberellic acid on zoysia grass. Left, treated; right, untreated

Fig. 8-5. Effect of ethephon in reducing development of last internode in winter barley to prevent lodging. Treated plants at left; untreated at right. Photograph courtesy of Union Carbide Agricultural Products Company

Fig. 8-6. Effect of chlormequat in reducing stem length in wheat allowing it to withstand severe lodging. Treated area in upper section; untreated in lower section. Photograph courtesy of American Cyanamid Company

a 

Fig. 8-7 a, b. Effects of mepiquat chloride on cotton. **a** Effect on vegetative plants.

Fig. 8-7b. Effect on fruiting structures. Photographs courtesy of BASF Wyandotte Corporation

Fig. 8-8. Shortening of cotton plants by use of plant growth regulators. Left: treated; right: untreated. Photograph courtesy of U.S. Department of Agriculture

Chapter 9. Axillary Buds

In tobacco production, decapitating the primary stalk is done in order to obtain better leaves with the desired physical properties and chemical composition. Unfortunately, this action removes the apical dominance of the primary stalk, allowing subsequent growth of axillary buds or "suckers". The formation of suckers in tobacco decreases leaf quality. The standard commercial practice for controlling axillary buds in tobacco has been the use of maleic hydrazide, which removed the necessity for large amounts of hand labor previously required in removing such buds. This use of maleic hydrazide is one of the great success stories in the history of plant growth regulators. Controversy over the safety of using this compound (see chapter on toxicology) led to a resurgence in research for other sucker control

a

b

Fig. 9-1. Prevention of sucker development in tobacco (flue-cured type) by treatment with maleic hydrazide. **a**: topped, but untreated (note abundant development of suckers); **b**: topped, then treated with maleic hydrazide (note absence of any sucker development). Photographs courtesy of U.S. Department of Agriculture

agents. The methyl esters of certain fatty acids also have been found to be effective for sucker control [477]. Outstanding among current cadidates are several nitroanilines and nitrophenylhydrazines [478–486], the most effective of these being currently in the process of commercialization [487–490].

Recent reports of active compounds for control of tobacco suckers include pyridazinones [491], diphenyl ethers [492], acetophenones [493], and dihydroimidazoisoindolediones [494, 495].

Camptothecin, the anti-tumor agent from the plant *Camptotheca acuminata* [496], completely inhibits growth of suckers in tobacco plants that have been topped [497].

Chapter 10. Chemical Pruning

Although chemical pruning falls under the general classification of controlling plant height, sufficient work has been done on this specialized use to consider it separately. Sachs and Hackett [498] described three basic methods for restricting plant height by using chemicals to modify stem growth:

(a) by selectively retarding internode elongation without interfering with the function of the apical meristem;
(b) by selectively reducing the activity of, or by killing, either the apical meristem or the terminal meristem of lateral branches; or
(c) by temporarily reducing apical dominance – thus encouraging simultaneous growth of lateral shoots.

Chemical pruning, often called "pinching" or "suckering", is done by the latter two methods. Feasibility of chemical pruning of plants was shown in the mid 1960's [499]. One of the first chemicals used was methyldecanoate [499–502]. Later, maleic hydrazide and the morphactins were commonly used. A more recently discovered compound is dikegulac sodium. First reported in 1975 [503] and since then the most widely used of the pinching agents, dikegulac sodium has been found to be effective on a large number of plants, particularly azaleas [504–512] and chrysanthemums [513], and also including begonia [514], sunflower [513], fuchsia [514], *Kalanchoe* [514], *Pachystachis* [514], *Zinnia* [513], *Euonymus* [515, 516]. In addition, it has been reported effective on woody ornamentals such as oleander, *Pyracantha, Callistemon, Xylosma,* Rhododendron, *Cotoneaster*, holly, *Pieris, Terstroemia* [517]. It has also been found useful for the pruning of young pecan [518] and walnut [519] trees. With pecan trees, treatment with dikegulac sodium in the autumn promoted substantial lateral shoot development on twigs the following spring [520]: mechanical pruning of untreated trees did not induce such lateral branching.

The manual removal of side shoots is an essential but time-consuming operation in commercial *tomato production*; it has been estimated that this operation amounts to about 5 percent of total production cost. In addition to the labor costs involved, manual pruning can result in injury to the plant and is a potential source of the spread of disease. Recent work in England has shown that methyl esters of the long-chained fatty acids and of long-chained aliphatic alcohols are effective in killing the side shoots of tomatoes, preventing regrowth for at least two months [521, 522, 523]. These same esters of fatty acids or fatty alcohols, as well as ethyl-5(4-chlorophenol)2H-tetrazole-2-yl-acetate, can be used on Brussels sprouts plants to prevent excessive plant height, to promote uniform sprout development along the stalk, and to hasten maturity [524, 525]. Previously, this removal of the terminal bud was a manual operation.

Most *poinsettia* cultivars display strong apical dominance, thus the terminal bud must be removed to produce a branched plant. To do this manually is a costly procedure and results in tissue loss and damage to the plant. Undecanol, which has been previously found to be effective in chemical pinching of chrysanthemums [526], azalea [506], is effective on poinsettia [527].

Surfactants such as alkarylpolyoxyethylene glycol-containing free fatty acids have been shown to destroy apical meristems in *chrysanthemums* giving them an excellent potential for chemical pruning [528].

Chapter 11. Plant Shape

Many important cultivars of apple and other fruit trees produce few or no branches on maiden trees in the nursery, yet an important requirement of a young fruit tree is that it be well branched. Young orchards planted with well-branched trees flower more profusely and crop much more heavily than those planted with poor quality, sparsely branched trees [529]. Plant growth regulators can be used in several ways to improve the shape of such trees. First, growth retardants can be used in the nursery on branches produced too low near the base of the tree. Second, treatment with chemical branching agents can improve the branching of those cultivars that normally form few branches in the nursery. Third, "chemical pruning" has potential uses on young fruit trees in the orchard and on ornamental nursery trees.

Several chemicals stimulate branching in apple and pear cultivars in the nursery [529–536]. These include 3-phenyl-1,2,4-thiadiazol-5-yl-thioacetic acid, propyl-3-*t*-butylphenoxy acetate, and a mixture of fatty acid esters. Propyl-3-*t*-butylphenoxy acetate has become so well known by its coded designation, M&B 25–105, that its code name has been registered as the trade name for this material [536]. Cherry and plum trees have also responded favorably to this chemical. Another compound found to be effective in stimulating branching in apple, pear, and plum trees is ethyl-5(4-chlorophenol)-2H-tetrazole-2-yl-acetate [537–538]. In addition to improving the shape of the young tree, this compound improves early flowering and early commercial cropping giving consistent yield increases. So far, results with sweet cherry trees have not been encouraging.

Chapter 12. Tillering

A tiller is a branch (stalk) from the base of the plant or from the axil of a lower leaf. Tillering is a general characteristic of grasses. The manner of tillering provides a means for dividing grasses into two groups: tufted grasses and sod formers. In sod formers, there is intense underground branching which permeates the soil. These underground branches combined with the root system forms a coherent mat which is found in lawns and in permanent pastures. In tufted grasses, the underground branching is limited. This is followed by the formation of a number of erect stalks. The individual plants (in clumps) are easily distinguishable. The cereals and sugarcane belong to the tufted grasses.

For many crops, the characteristic most related to yield is the number of healthy stalks per unit area at harvest [539]. Because of this, inducing tillering is often the most effective way to increase yields in a number of crops.

Leopold [540] found that tillering in barley was stimulated by spraying with *2,3,5-triiodo-benzoic acid* (TIBA) – a finding confirmed by other investigators [541]. Langer and his coworkers found that TIBA had a similar effect on wheat [542]. Cytokinins such as *6-benzylamino-purine* (BAP) and *6-benzylamino-9-(tetrahydropyran-2-yl)-9H-purine* (PBA) also have been shown to increase tillering in wheat and barley [542, 543].

Work in India with a dwarf wheat ("Hira") showed triiodobenzoic acid to increase tillering and ear number in this dwarf *aestivum* variety but to have little effect on a semi-dwarf variety "Kalyansona". Gibberellic acid did not affect the tiller number in "Hira" but reduced that of "Kalyansona"; it also delayed tillering and improved tiller survival in both varieties [544]. Work with rice in Taiwan showed that the auxin transport inhibitors, DPX-1840 and chlorflurenol, promote tillering [545].

Recent results from Jamaica show sugarcane "seed pieces" (setts) to respond to ethephon [546]. Two of the three cultivars studied tillered freely in response to ethephon treatment; the third did not tiller – regardless of treatment. These results confirm the work with ethephon previously conducted in Hawaii [547]. Results from Brazil show no effect of ethephon in the plant crop but a significant increase in the number of millable stalks in the ratoon [548].

Chapter 13. Resistance to, and Control of, Insects and Diseases

Insects

Using plant growth regulators for the control of different insect pests has been rekindled recently, although potential use of these substances for this purpose goes back several decades. One of the first pieces of work was conducted by the United States Department of Agriculture in Louisiana to evaluate the effect of 2,4-D, used in sugarcane fields for weed control, on the sugarcane borer [549]. In comparing those plots that received 2,4-D only with those that received no treatment, an increase in borer population was found for the plants treated with 2,4-D. It was concluded that the heavier infestation in the treated plots in this experiment was due largely to the detrimental effect of the 2,4-D treatment on the parasitism of the sugarcane borer. The direct spraying of 2,4-D amine on coccinellid larvae increased mortality and slowed the development of the living larvae to the pupal stage [550].

Aphids are among the group of plant-feeding insects closely adapted to their hosts; consequently, even slight variations in the growth and physiology of the host plants might be expected to have significant effects on the aphids feeding on them. Robinson in Canada worked with thirty plant growth regulators and herbicides including gibberellic acid, monuron, 2,4-D, 2,3,6-trichlorobenzoic acid, and maleic hydrazide, to determine their effect on the pea aphid living on the broad bean. Of the compounds tested, only three were found to decrease both nymph mortality and fecundity of the female aphid; of these three, one was the plant growth regulator maleic hydrazide [551, 552, 553]. In these studies Robinson found no effect of 2,4-D on adult or nymphal mortality, nor on adult fecundity; whereas Maxwell and Harwood had found the reproductive rate of pea aphids to increase on broad bean plants treated with 2,4-D [554, 555]. Later, Van Emden [556] working with cabbage aphids on Brussels sprouts in England found reduced progeny on chlormequat-treated plants. He concluded that the (a) poor longevity, (b) decreased fecundity, and (c) depressed rate of increase on the treated plants is probably a nutritional effect rather than the effect of toxic properties of chlormequat itself. His results led Van Emden to suggest that if this observed resistance to the aphid in treated Brussels sprout plants holds in other host plant/aphid combinations, growth retardants may have a very useful secondary property for the pot plant industry, particularly under glass.

Work at the University of Reading with plant growth stimulators and with plant growth retardants showed an alteration of the fecundity of the aphids [557]. The reduced fecundity of aphids on broad bean treated with gibberellic acid is thought to be related to the apparent toxicity of the compound when fed directly to the aphids. Chlormequat seemed toxic to both types of aphids used in the study. Apart from any direct action of chlormequat on the aphids, there was evidence that the substance may reduce the availability of nutrients from the treated plants, thus resulting in an adverse effect.

Studies with two-spotted spider mites raised on snap beans showed that treatment with gibberellic acid resulted in significant reductions in mite populations [558]. Similar effects were found on gibberellin-treated beans and apple trees, whereas gibberellin-treated cotton plants gave variable results [559].

Work in Israel showed that chlorphonium, carvadan, and chlormequat had a pronounced anti-feeding effect on the cotton leafworm feeding on wax bean plants growing in pots; foliage application was more effective than soil treatment [560]. Work by the Anti-Locust Research Center in England showed the importance of gibberellin in hastening ecdysis in the desert locust [561, 562]. Feeding experiments showed that a diet of senescent leaves delayed sexual maturation in the desert locust; the senescent leaves were shown to be short of gibberellin. The authors suggested that the sexual immaturity of desert locust during the dry season may result from the senescent condition of their desert food plants. This was confirmed in subsequent work in which the same investigators showed that gibberellic acid is a necessary dietary constituent for normal, fast maturation in the desert locust [563]. They also found that chlormequat acts directly upon the insects, apart from any effect produced by the altered physiology of the food plant.

A recent study was conducted in Egypt to evaluate plant growth regulators for their ability to impair the growth of, or to serve as anti-feedants for, insect larvae (greasy cutworm) [564]. In this study, the test compounds were injected. Chlormequat caused the formation of small-sized pupae that gave rise to relatively small adults. Daminozide had a juvenilizing effect resulting in the formation of giant super-larvae. Although they consumed as much food as the controls, 2,4-D-treated larvae had low body weights. Work in the same laboratory showed the anti-feedant properties of naptalam and amitrole on the American bollworm [565].

In 1958, Adkisson [566] showed that populations of the cotton pink bollworm generally decreased in fields following applications of desiccants or defoliants. Further research indicated that the days in central Texas are not short enough to induce a large percentage of the insect population into diapause until mid-September [567]. These findings suggested that using chemicals to terminate growth of cotton plants before the days are short enough to induce diapause in the pink bollworm might be an effective method for reducing the number of diapause individuals in the population [568]. This work resulted in a series of investigations that extend to the present time evaluating plant growth regulators for the control of this cotton pest [569–578]. 2,4-D, chlormequat, and chlorflurenol proved to be effective for this purpose, as have two other regulators – N-dimethyl morpholine chloride and 2,3-dichloroisothiazole-5-carboxylic acid. Best results appear to be obtained by using mixtures of two growth regulators – one that is fast acting but not persistent and the other that is slow acting but persistent [575]. Chlorflurenol and chlormequat are the persistent growth regulators. The fast-acting ones include 2,4-D and 3,4-dichloroisothiazole-5-carboxylic acid. Results reported in 1976 suggested that both the pink bollworm and the boll weevil populations would be deprived of the food supply for the development of the diapause generation by treatment with plant growth regulators. More recently this work has spread to evaluate the suppression of feeding sites for populations of the bollworm and the tobacco budworm [579]. Results of such investigations indicate that plant growth suppressants can be used effectively to reduce late season vegetative and reproductive growth and to make major contributions to bollworm-budworm control.

The most recent consensus is that, when properly timed, chemical termination of growth in cotton will reduce by about 95 percent the number of bolls that will not mature and, because of the reduced food supply, will reduce the over-wintering generation of pink bollworm by about 95 percent. For reducing pink bollworm populations, chemical termination was found to be superior to defoliation, desiccation, or early irrigation termination and is felt to be adaptable to integrated pest management programs.

Diseases

The concept of disease control through induced changes in the metabolism of the plant host by the application of growth regulators has continually been studied since Davis and Diamond in 1953 [580] showed that 2,4-D, NAA, TIBA, NOA, and IAA reduced Fusarium wilt disease in tomatoes. Sinha and Wood [581] confirmed positive results with IAA, chlormequat, and NAD, mixed results with 2,4-D and TIBA, and an increase in disease in gibberellic acid-treated plants.

Studies with charcoal rot disease of soybeans showed variable results with IAA and kinetin. However, gibberellic acid and TIBA reduced disease severity both under greenhouse and field conditions [582].

There are conflicting reports as to whether exogenously supplied cytokinins increase or decrease virus production in infected plant tissue. Several investigators have found increases in virus production [583, 584, 585], some report decreases [586, 587], and some have obtained variable results depending upon a number of factors such as time of application, concentrations used, and age and condition of the host plants [588–591].

Although Verticillium wilt of cotton can be partially controlled by using resistant cultivars, a high degree of resistance is lacking; therefore, other methods of control are necessary. Results with systemic fungicides, such as benomyl, have been only partially effective and not economical. Recent investigations in California show that plant growth retardants, when applied to cotton, delay the onset and slightly mitigate the severity of Verticillium wilt symptoms [592–595]. Such treatment also reduced the internal population of the disease in the plants. The effective growth retardants included chlormequat, N,N-dimethyl piperidinium chloride, and tributyl (5-chloro-2-thienyl)methyl phosphonium chloride.

Results from Rothamsted Experimental Station in England show that single foliar sprays of daminozide approximately halved the incident of common scab in potted potato plants in the greenhouse [596]. Two analogs of daminozide were active, the remaining ones were inactive. Of the twenty-two other unrelated plant growth regulators evaluated only gibberellic acid decreased scab incidence, but many of the tubers were distorted. The conclusion of the investigators is that daminozide decreases the incidence of scab by altering the physiology of the plants so that the symptoms do not develop.

Studies in Russia show that treatment of wheat with chlormequat decreased cercosporiosis by control of the lodging [597]. Chlormequat also lowers virus muliplication in tobacco [598].

Chapter 14. Overcoming Environmental Stress

Over twenty years ago it was observed that plants treated with various growth retardants were less susceptible to external stress conditions such as water [599], temperature, salinity [600], diseases [598], and pests [556].

Water

Early work in Israel with growth retardants – including chlormequat, phosfon, phosfon S, and daminozide – showed that all of the chemicals tested caused an increase in the dry weight of plant roots [601]. Chlormequat reduced the transpiration rate per unit of leaf area whereas the other chemicals in this test either increased or had no effect on transpiration. The investigators concluded that the more extensive root system of the treated plants and their lower top/root ratios contribute to their survival under conditions of water stress, specially under field conditions. They deduced that these growth-retarding substances probably increase drought avoidance even though they have no effect on the drought tolerance of plants. Recent preplant soil incorporation studies in Russia with chlormequat for peas showed increased drought resistance and increased yields [602].

Plant hormones were implicated early in studies of the regulation of water balance: cytokinins resulted in increased transpiration by causing the opening of stomata [603–607]; abscisic acid caused the reverse effect – the closing of stomata and a resultant reduction in transpiration [607–615].

The use of anti-transpirants to reduce water loss in plants has been extensively studied. Anti-transpirants may be of the metabolic type which induce the closure of stomata, or those that coat the leaf surfaces and act as inert, waterproof films. Anti-transpirants of the metabolic type can be represented by phenylmercuric acetate [616–624], by the naturally-occurring hor.one abscisic acid [607–615], or by others (see Table 14–1 for list of metabolic anti-transpirants). Materials such as latex [635] and silicone [614] can be used to coat the leaf surfaces to prevent transpiration. Thus we have three ways to reduce plant transpiration: a) physical covering of the stomata, b) chemicals that cause a closure of the stomata, and c) compounds such as the growth retardants that reduce the water stress by reducing plant growth. Excellent reviews of plant anti-transpirants are to be found in references [624, 636, 637].

Table 14–1. Chemicals as Metabolic Anti-Transpirants. (After Das and Raghavendra, 1979, p. 624)

Compound	Test Plant	Reference
Abscisic acid	ash	614
	barley	609, 610, 625
	bean	626
	citrus	613
	cucumber	626
	peppers	626
	pine	614
	sugar maple	613
	tomato	626
	Xanthium strumarium	627
Alachlor	corn	628
Alkenylsuccinic acid	tobacco	629
Chlormequat	sunflower	630
	tomato	631
2-Chloromercuri-4,6-dinitrophenol	*Datura arborea*	632
Daminozide	tomato	631
2,4-Dinitrophenol	*Datura arborea*	632
8-Hydroxyquinoline	tomato	631
	strawberry	633
Indoleacetic acid	tomato	626
Chlorflurenol, methyl ester	cotton	634
Phenylmercuric acetate	birch	623
	cotton	620
	Festuca rubra	622
	tobacco	617, 618
	tomato	631
Salicylaldoxime	*Datura arborea*	632

Recent work has led to attempts to discover the ideal chemical anti-transpirant for use on crops in dry regions [636, 637]. However, most of the known chemical anti-transpirants have undesirable toxicological side effects both for plants and for animals [628]. Recent work in India showed that the herbicide alachlor could induce partial stomatal closure in corn [628]. Treatment with alachlor produced higher rates of photosynthesis as well as increased growth rate and yield; plants in the same test treated with phenylmercuric acetate exhibited the reverse trend.

Work in Mexico with drought-resistant and susceptible strains of both sorghum and maize points out that there is an alternative for the grower to use either a drought-resistant cultivar or use a susceptible one treated with chlormequat. However, they found no advantage from the treatment of drought-resistant cultivars with chlormequat [638–639].

Salt Tolerance

In 1961, U. S. Department of Agriculture investigators reported increased tolerance of plants to toxic levels of salts after treatment with growth retardants such as Amo-1618, phosfon, or chlormequat [600].

Temperature

In his review of the physiology of growth retarding chemicals, Cathey [640] reviewed the various effects of growth retardants, including their effect on the resistance of plants to frost injury. Since that time a number of investigators have studied the effects of growth retardants on the ability of plant systems to withstand changes in temperature. Marth [641] found that frost damage to cabbage plants was markedly reduced by the application of chlormequat or daminozide prior to exposure of the plants to low temperature. Wunsche [642] showed a definite increase in frost hardiness of winter wheat following applications of chlormequat. Similar results were obtained with tomato plants [643]. Subsequently, daminozide was shown to increase the cold hardiness of box elder [644] and raspberry [645].

Working with cucumber seedlings, Rikin and his coworkers [646, 647] concluded that abscisic acid is involved in the development of chilling resistance in these plants. The abscisic acid can either be applied to the seedlings or be induced to increase endogenously by exposure to water shortage. Similar results were subsequently shown for cotton [648].

Investigators at the Asian Vegetable Research and Development Center in Taiwan found that Chinese cabbage seeds treated with a mixture of gibberellins A4 and A7 for 24 hours before placing them on a germination medium provide the selection criterion to identify heat-tolerant varieties. This finding has been developed into a rapid and useful technique to screen for heat tolerance in Chinese cabbage at the seedling stage [649].

2-Amino-6-methylbenzoic acid at rates of about 1 lb/acre increases frost resistance under laboratory conditions in wheat, tobacco, and grape plants [377]. Polyamine compounds, such as long-chained alkylene diamines, have been shown to protect a number of crops from cold weather or frost damage [650]. These crops included wheat, rice, barley, oats, corn, millet, soybean, lima beans, pole beans, snap beans, navy beans, peanuts, spinach, lettuce, tomato, mulberry, tobacco, several fruit trees, and a number of flowering plants including poinsettia, carnation, and geranium.

5-Chloro-4-quinoline carboxylic acid, 2-chloro-4-quinoline carboxylic acid, and 2-trifluoromethylquinoline carboxylic acid have been shown to increase significantly the number of surviving seedlings of zucchini squash subjected to a 24-hour freezing cycle [204].

Water spot is the term applied to a certain type of breakdown of the rind of the navel orange and is an important economic factor in the production of this crop. The condition is non-pathogenic and develops on the surface of a mature or nearly mature navel orange that has been wet with water for several days, a condition not uncommon in the field under autumn conditions. Aqueous sprays of gibberellic acid at low concentrations are effective in alleviating this condition [651].

Chapter 15. Mineral Uptake

A study of the influence of gibberellic acid, 2,4-D, and daminozide on the uptake of nitrogen, phosphorus, and potassium by wheat and soybean plants showed that uptake of potassium by wheat is accelerated by gibberellic acid; 2,4-D increased the uptake of nitrogen and phosphorus in both wheat and soybean plants [652]. Both gibberellic acid and 2,4-D interfered with the translocation of all three minerals. Daminozide did not affect the uptake of any of the three minerals but did stimulate the translocation of phosphorus and potassium within the plant. Studies in Brazil with tomato plants showed that foliar applications of chlormequat resulted in higher concentrations of nitrogen, calcium, and magnesium in the stems of treated plants. Similar treatment with daminozide caused an increase in the nitrogen concentration of the stems.

Russian studies [653] on winter wheat in the field show that chlormequat decreased the nitrogen and retarded protein accumulation; on the other hand, spraying with a combination of chlormequat and the amine salt of 2,4-D increased the nitrogen concentration and stimulated protein accumulation. Studies with okra in India [654] showed that foliar application of chlormequat or daminozide reduced the accumulation of both nitrogen and phosphorus in the fruit. Tomato plants treated with chlormequat had higher nitrogen, calcium, and magnesium concentrations than untreated plants [655, 656]; those treated with daminozide had increased nitrogen concentrations [656].

Chapter 16. Plant Composition

To the vocabulary of the chemical regulation of plant growth by compounds under such names as hormones, growth regulators, herbicides, etc., another term, *bioregulator*, has been introduced by Meier and Yokoyama [657]. Their definition of the term bioregulator denotes substances that control specific biosynthetic or metabolic pathways, or steps of pathways, without limitation as to whether the pathways are specifically essential to growth. Their key point is that a hormone has a broad physiological impact on the plant and this manifests itself as a gross morphological effect in terms of growth or development. On the other hand, their concept is that a bioregulator has a narrow biochemical effect that results in a specific change in the composition of the plant that may not be observable as a morphological effect. They point out as a good illustration the effect of a bioregulator on the pathway of secondary metabolism for pigment accumulation. The bulk of their work has been with citrus.

Citrus Color

Fruit color is of great economic importance to the citrus industry because consumers prefer oranges with a deep orange color over those having a yellow or a yellow-orange color. The most desirable citrus color, as visualized by the consumer, is essentially due to the presence of a class of pigments known as the carotenoids. Nevertheless, a large portion of the orange crop does not possess the most desirable coloration so there is a strong commercial interest in obtaining the desired color, preferably through natural means – particularly in view of the growing concern over the safety of artificial colors. Earlier work showed that hormones can affect color changes in the peel of citrus. Such hormonal effects are due to a general decrease in carotenoid content rather than a specific stimulation and accumulation of individual carotenoid pigments [658].

A major step forward was achieved in 1970 by the discovery that a chemical agent, *2-(4-chlorophenylthio)triethylamine hydrochloride* (CPTA), can cause the formation of and accumulation of the intensely red carotenoid pigment lycopene in plants, particularly in citrus fruits [659]. It is interesting that, although Knypl [660] found that chlormequat causes the accumulation of lycopene in pumpkin cotyledons, no such response is obtained when chlormequat is applied to citrus fruits [659]. Following the discovery of the effect of CPTA, considerable effort has been devoted to the control of pigmentation in citrus fruits – with very successful results. Three groups of compounds have been found to be effective in citrus coloration. The first group, which includes CPTA and other triethylamine derivatives, causes rapid and intense accumulation of the red pigment lycopene and, to a lesser extent, γ-carotene [661, 662]. Members of the second group, which includes N,N-diethylnonylamine, cause the production of small amounts of three carotenoids (α-, β-, and γ-carotenes), all of which are orange, as well as lycopene [663, 664]. The third group, composed of representatives of a newer group of amine derivatives, comes closest to meeting the needs of the citrus industry; these compounds produce a substantial increase in orange-colored carotenoids, but only small amounts of lycopene. Because of the lower lycopene content, fruit color developed usually never goes beyond deep orange.

Whereas the final color of most orange cultivars is produced by a decline in chlorophyll pigments and an accumulation of carotenoids, the final color in most lemon, lime, and grapefruit cultivars is produced by a decline in chlorophyll pigments with little or no net increase in carotenoids [665]. In addition to the work of Yokoyama and his colleagues discussed earlier in this chapter, considerable information is available concerning other growth regulators and their effects on the coloration of citrus fruits. For example, gibberellic acid causes a substantial delay in the loss of chlorophyll and a substantial reduction in the rate of carotenoid accumulation. 2,4-D and benzyladenine cause delays in loss of chlorophyll but have little or no influence on carotenoids. In the light, another compound, 2',4'-dichloro-1-cyanoethanesulfonanilide, causes a rapid and uniform loss of chlorophyll; in darkness, it does not have a degreening effect but interferes with ethylene degreening, suggesting that its mode of action is different from that of ethylene [666, 667].

Color in Other Fruits and Vegetables

Ethephon has been shown to increase red coloration in fruits of tomatoes [668, 669], bell peppers [670–673], chili peppers [674] and pimiento peppers [675], cranberries [676–679], certain table grape varieties [680, 681], and many apple cultivars [682–689]. Daminozide has been shown to be effective in increasing the red coloration of sweet cherries [690–694].

Essential Oils

Recent results from Florida show that when "Hamlin", "Pineapple", and "Valencia" orange cultivars were treated with abscission agents to loosen the fruit to facilitate mechanical harvesting, six phenolic ethers were isolated and identified for the first time from the treated fruit. None of these ethers was found in the non-treated fruit. The compounds isolated and identified for the first time as citrus constituents were eugenol, methyleugenol, cis-methylisoeugenol, trans-methylisoeugenol, elemicin, and isoelemicin [695].

Tobacco

Ethephon is effective as a "yellowing" agent for assisting in the ripening and curing of flue-cured tobacco. Results from Virginia show that, when used for this purpose, ethephon causes a significant increase in reducing sugars and a decrease in total nitrogen in the leaves of treated plants. Other chemical constituents, including nicotine and petroleum ether extractables, were not affected [696].

Fig. 16-1. Ethephon field-treated tobacco after curing. Treated leaves at left; untreated at right. Photograph courtesy of Union Carbide Agricultural Products Company

Chapter 17. Metabolic Effects, Ripening, and Yield Increases

Compounds that affect crop metabolism, particularly those that regulate crop maturity, are especially likely to have a dramatic impact on agriculture in the years ahead. Several such compounds already are used commercially and their success is in large part responsible for increased interest in plant growth regulators in agriculture. Most of the compounds used on economic crops have a direct or indirect effect on final yield, on quality, or on both.

Sugarcane Ripeners

One of the most important developments in recent years has been the use of chemicals as ripeners on sugarcane. Although attempts were made for several decades to control the ripening of sugarcane by the use of chemicals, no concerted effort was made until the start of a research program in Hawaii in the early 1960's [131, 132, 698]. This effort was soon joined by investigators in Australia [699–702] and Trinidad [703, 704]. The initial success resulted in extensive field testing throughout the sugarcane world [705], especially in Mauritius [706–710], South Africa [711–715], India [716–721], Brazil [722, 723], the Philippines [724, 725, 726], Taiwan [727–730], Guyana [731, 732], Colombia [733], Puerto Rico [734–736], and the mainland USA [737, 738, 739]. Originally, very few commercial companies were involved; chemicals used were primarily those available from chemical supply houses together with the few materials synthesized by research organizations in the sugarcane industry around the globe. The initial success led a number of companies to become interested in supplying chemicals for evaluation. This total effort has resulted in a surprising number of chemicals that increase the sucrose content of sugarcane at harvest. Most of those compounds that have met with sufficient success to have information published about them are given in Table 17–1. Some compounds were never developed beyond the initial screening stages; others are too new to have been reported other than through an initial publication or an issued patent.

The first material seriously considered as a candidate ripener for increasing sucrose yields of sugarcane was the dimethylamine salt of 2,3,6-trichlorobenzoic acid. Because of a number of technical, environmental, and legal problems, this material did not prove successful commercially [760]. Nev-

Table 17–1. Sugarcane Ripening Compounds

Compound	Common Name	Code Designation	Trade Name	Source	Reference
p-aminobenzenesulfonyl urea	–	–	–	Monsanto	740
2-amino-6-methyl benzoic acid	6-amino-o-toluic acid	ACR-1308	–	Maag	443, 741
aminomethylphosphonic acid	–	AMPA	–	Procter & Gamble	742
6-aminopenicillanic acid	–	6-APA	–	Various	743, 744
4-amino-3,5,6-trichloropicolinic acid	picloram	–	Tordon	Dow	698, 744
ammonium ethyl carbamoyl phosphonate	–	–	–	duPont	745
ammonium isobutyrate	ammonium isobutyrate	AIB	–	Various	746
N-benzoyl-N-(3,4-dichlorophenyl)-amino-propionic acid	–	–	–	Shell	747
N,N-bis(phosphonomethyl)glycine	glyphosine	CP-41845	Polaris	Monsanto	131, 132, 699, 701, 702, 707–710, 717–723, 725, 726, 728, 729, 734, 736–739, 748–756
bis (N,O-trifluoroacetyl)-N-phosphonomethyl glycine	–	–	–	Monsanto	757
5-bromo-3-*sec*-butyl-6-methyluracil	bromacil	–	Hyvar X	duPont	758
2-chlorobenzoic acid	–	–	–	Various	131, 744
2-chloroethylaminodi(methyl phosphonic acid)	–	–	–	Monsanto	759
2-chloroethylphosphonic acid	ethephon	Amchem 66-329	Ethrel Cepha	Union Carbide GAF	131, 132, 701, 712–715, 717, 725, 729, 733–735, 744, 749, 750

Table 17–1. (continued)

Compound	Common Name	Code Designation	Trade Name	Source	Reference
2-chloroethyltrimethyl ammonium chloride	chlormequat	CCC	Cycocel	American Cyanamid	132, 698, 710, 716, 717, 720, 734, 744, 750–752, 760–763
5-chloro-2-thenyl-tri-n-butylphosphonium chloride	–	CHE-8728	–	Chemagro	133, 744, 750
3-cyclohexene-1-carboxylic acid	tetrahydrobenzoic acid	–	–	Various	744
2,3-dichloro-6-methylbenzoic acid	–	–	–	duPont	764
N-(2,3-dihydroxy-1-propyl)-N-phosphonomethyl glycine, disodium salt	–	–	–	Monsanto	765
Diisobutylphenoxyethoxy-ethyl-dimethylbenzyl ammonium chloride	–	–	Hyamine 1622	Rohm & Haas	131, 718, 721, 729, 744, 750–752, 766
2-(β-dimethylamino-ethoxy)-4-(3′,4′-dichlorophenyl)-thiazole hydrochloride	–	–	–	Ciba-Geigy	767
dimethylarsenic acid	cacodylic acid	–	Phytar 138	Ansul	131, 744, 768
N,N-dimethylglycine	–	–	–	Monsanto	769
3-(2-[3,5-dimethyl-2-oxocyclohexyl]-2-hydroxyethyl)glutarimide	cycloheximide	–	Actidione	UpJohn	131, 744, 762
N-[2,4-dimethyl-5-[[(trifluoromethyl)sulfonyl]-amino]phenyl]acetamide	mefluidide	MBR-12325	Embark	3M	131, 132, 458, 725, 729, 744
Hexadecyltrimethylammonium bromide	cetyltrimethyl ammonium bromide	CTAB	Cetrimide	Various	131, 718, 721, 744, 750–752, 770
1-hydroxy-1,1-ethane diphosphonic acid	–	–	–	Monsanto	771

Table 17–1. (continued)

Compound	Common Name	Code Designation	Trade Name	Source	Reference
4-hydroxy-3-methoxybenzaldehyde	vanillin	–	–	Ontario Paper Company	131, 744, 772
imidodicarbonic diamide	carbamyl urea	–	Biuret	Nipak	773
isochlortetracycline	isoaureomycin	–	–	American Cyanamid	131, 744, 750, 760, 762, 774, 775
laurylmercaptotetrahydropyrimidine	–	–	–	Pfizer	131, 698, 744, 750, 760, 762, 776
β-mercaptovaline	penicillamine	–	Cuprimine	Merck	743, 744
methyl-3,6-dichloro-o-anisate	disugran	60-CS-16	Racuza	Velsicol	131, 717, 725, 735, 736, 744, 750, 751, 777
7-methyl indole	–	PP-757	–	ICI	729, 778
3-(2-methylphenoxy)pyridazine	–	H-722	Credazine	Sankyo	131, 744, 750, 779
2-methyl-1-propanol	isobutanol	–	–	Various	780
methylsulfanil-yl-carbamate	asulam	MB-9057	Asulox	May & Baker	131, 729, 744
2-(p-methoxybenzyl)3,4-pyrrolidine-diol-3-acetate	anisomycin	–	Flagecidin	Pfizer	131, 744, 762
N-[(4-methoxy-6-methylamino-1,3,5-triazin-2-yl)-aminocarbonyl]benzene sulfonamide	–	–	–	duPont	781
7-oxabicyclo-(2,2,1)-heptane-2,3-dicarboxylic acid, monoalkylamine salt	Endothall, monoalkylamine salt	TD-191	Ripenthol	PennWalt	132, 698, 744, 750–752, 760, 762, 782
n-pentanoic acid	n-valeric acid	–	–	Various	131, 744, 783
6-phenoxyacetamido-penicillanic acid	penicillin V	Pen-V	Pen-Vee	Various	131, 744, 784
N-(2-phenoxyethyl)-N-propyl-1H-imidazole-1-carboxamide	–	BTS 34-273	–	Boots	785, 786

Table 17–1. (continued)

Compound	Common Name	Code Designation	Trade Name	Source	Reference
N-phenylphosphinylmethyliminodiacetic acid-N-oxide	–	–	–	Monsanto	787
N-phenylsulfonamido-N-phosphonomethyl glycine	–	–	–	Monsanto	788
Phosphonic acid, (2,2,2-trichloro-1-hydroxyethyl)-bis-[2-(2-hydroxypropoxy)-1-methylethyl]ester	–	–	–	American Cyanamid	789
N-phosphonomethylglycine	glyphosate	MON-8000	Polado	Monsanto	131, 132, 700, 701, 715, 726, 739, 744, 750, 751, 753, 756, 790
poly[oxyethylene(dimethylimino)ethylene(dimethylimino)-ethylene dichloride]	–	–	Bualta	Buckman	701, 725, 729, 791, 792
tetrahydrofuroic acid hydrazide	–	–	–	Various	131, 744, 793
1,2,4-triazine-3,5(2H,4H)-dione	6-azauracil	–	–	Various	131, 744
N-trichloroacetylamino methylenephosphonic acid	–	–	–	Monsanto	794
2,3,6-trichlorobenzoic acid, dimethylamine salt	2,3,6-TBA	TBA	Trysben	duPont	131, 132, 698, 703, 725, 736, 744, 760, 762, 775, 795
3-(trifluoromethyl-sulfonamido)-p-acetotoluidide	fluoridamid	MBR-6033	Sustar	3M	131, 717, 725, 744, 750–752
–	bacitracin	–	Bacitracin	Various	131, 744, 796
–	mineral oil	–	–	Esso	703, 797
–	–	–	Tergitol NPX	Union Carbide	131, 744, 751, 798
–	–	–	Tween-20	ICI	131, 744, 751, 799

ertheless, it served as a standard for comparison in screening tests aimed at finding better sugarcane ripeners. It continued to be the standard for comparison until the registration of the first ripener for sugarcane in the United States. This compound is N,N-bis (phosphonomethyl) glycine, known generically as glyphosine and marketed by Monsanto as the product Polaris [748, 800]. Until late 1980, this was the only compound registered for this use in the United States. In the fall of 1980, phosphonomethyl glycine was also registered as a ripener for sugarcane. This compound, known generically as glyphosate [753, 790], is marketed by Monsanto as the product Polado; it is the sodium salt of the same compound that is the active ingredient of the herbicide Round-Up.

The ethylene-producing compound ethephon is used commercially in sugarcane in South Africa [712, 713, 715] and Rhodesia [714]. The effectiveness of ethephon as a sugarcane ripener has not been comparable to that of glyphosine in some areas of the world, although its growth effects enable it to result in an increase in yield. These effects are being evaluated in research programs in the sugar industry. Several other compounds have been registered for experimental use as ripeners on sugarcane in the United States and are under test in other countries as well; these include chlormequat, mefluidide, and ripenthol.

The number, and very chemical nature, of compounds that are active as sugarcane ripeners suggests that there are several modes of action to enhance the ripening of sugarcane and that the active compounds might fall into any of these several classifications. There are also varietal differences, as well as differences due to

(a) the fertilizer status (particularly nitrogen),
(b) the age of the crop and its condition,
(c) the climate (both during the growth of the crop and prior to harvest),
(d) the physiological state of the cane, and
(e) the purity of the juice in the young growing tops.

These variables, and probably many others, suggest that there is room for a number of ripeners on sugarcane. Additional variables to be considered are

(a) phytotoxicity of the ripeners,
(b) the cost effectiveness of the compound under consideration, and
(c) the effects on the processing of sugarcane.

The effect on growth of subsequent ratoons is an extremely important consideraction for registration as well as long-term use of any ripener.

Probably because glyphosine was the first registered sugarcane ripener, the use of this compound had a meteoric rise, at least in the sugar areas of the United States. From its registration in 1972 and experimental use on a few hundred acres, it reached over 60,000 acres in Hawaii by 1977. A similar situation occurred in Florida, increasing from 178 acres in 1972 to more than 46,000 acres in 1979 [801]. Because of the greater activity and considerably lower costs with glyphosate, it is expected that glyphosine will be replaced very quickly. Recommended dosage for glyphosine is about 4 lbs. active per acre whereas for glyphosate it is about 0.5 lb. active per acre.

In South Africa and Rhodesia, glyphosine was not as effective as ethephon; consequently, ethephon was the most widely used ripener in that area. Glyphosate has been found to be quite active in the sugarcane-growing regions of southern Africa and it is expected that it will be an effective competitor for ethephon.

Interest seems to be declining for three materials that received early consideration: chlormequat, disugran and ripenthol. In several major sugar growing areas, mefluidide appears to be losing popularity, but it is being more seriously evaluated in tropical countries such as the Philippines.

The financial return to the grower is substantial through the use of ripeners; increased sugar yield produced by such compounds can be as much as 20 percent, depending on the variety of sugarcane treated as well as on prevailing weather and soil conditions. Chemical control of maturation in sugarcane is now an established practice. In fact, it is so established that many research organizations in the sugar industry are shifting part of their efforts to looking at other stages in the development of the sugarcane crop for additional potential uses and times of chemical treatment. It is thought ethephon might fall into this sort of economic use, not unlike that found for gibberellic acid.

Rubber – Latex Flow

The production of natural rubber since 1945 has never matched world demand. Consequently, synthetic rubber has been needed to meet the total demand for rubber. During the same period, burgeoning of the world petrochemical industry on the basis of cheap and abundant feedstocks has enabled synthetic rubber producers to take advantage of lower unit costs in filling the gap between natural rubber supply and overall demand.

The advent of the energy/material crisis, plus the global concern for environmental pollution, adds force to the general need for more natural rubber. The contrast is obvious between natural rubber, a renewable material produced by essentially non-polluting industry, and the synthetics, made from increasingly more scarce and expensive feedstocks. The basic economics of the situation have precipitated a sharp swing away from synthetic to natural materials. Production of latex concentrate is currently estimated at about 281,000 dry metric tons and over three quarters of this production is accounted for by Malaysia. A shortfall of both natural and synthetic rubber is the forecast.

The rubber tree *Hevea brasiliensis* is the primary production unit of the natural rubber industry. Changes in productivity of this plant can be brought about by manipulation of a number of factors including genetic, physiological, agronomic, horticultural, etc.; over half a century of organized applied research has resulted in a spectrum of innovations that are at different levels of implementation. One of these innovations is the use of plant growth regulators to enhance the flow of latex from the rubber tree. Results of early investigations on this process resulted in the use of 2,4-D, 2,4,5-T, or NAA as stimulants for latex flow [802–807]. Work since that time with many chemicals [808] has resulted in the use of ethephon applied to the tree bark, enabling the trees to express their full genetic potential by reducing or removing the physical barriers to flow [807, 809–813]; the amount of yield increases achieved in this manner can range from 60 to 100 percent. The use of ethephon has now become an estate practice in most rubber-producing countries [814–819]. Large-scale field experiments conducted to assess the effect of ethephon on rubber yield and profit under different systems of tapping have also been carried out in India; yield increases ranged from 36 to 130 percent, with the profit returns being even higher [820].

Ethephon is applied to a scraped area of the bark or tapping groove. This chemical releases the gas ethylene, a plant hormone, which brings about enhanced latex flow by delaying the plugging processes of latex vessels. The gain in economic profits is primarily through the use of shorter tapping cuts, reduced tapping frequencies, and the resultant reduction in labor costs. More recent research suggests that it might be feasible to obtain good yields simply by puncture-tapping, thus prolonging the potential life of the trees and substantially reducing capital investment [821].

The effective use of ethephon has spread throughout the rubber producing world from the research results reported in Malaysia.

Success in controlling latex flow has led to investigations of the use of plant growth regulators in many aspects of rubber production. Probably most important are the efforts to use plant growth regulators to reduce the immature phase of the rubber tree. Other research includes evaluation of root promoting compounds, the use of growth promoters in the propagation of cuttings, the use of regulators for the development of lateral roots in budded stumps, and the use of plant hormones for effective branch induction [822].

Papaya – Latex Flow

An interesting offshoot of the work with rubber is its transfer to the papaya plant. Early reports show that spraying papaya plants with ethephon significantly increased the flow of latex in treated plants [823]. This led to more recent studies on the effect of ethephon on the production of the proteolytic enzyme papain which is concentrated in the latex of the immature papaya fruit [824]. Application of ethephon resulted in a four-fold increase in papain yield, improved quality of the enzyme, and improved quality of the fruit by increasing the total and reducing-sugars and by decreasing acidity.

Guayule – Latex Production

Parthenium argentatum, the guayule plant, is a low shrub native to the arid regions of north central Mexico and the Big Bend area of western Texas. The stems and roots of this plant contain a rubber that, when purified, is virtually the same as the natural rubber from the familiar rubber tree (*Hevea*). During World War II, when supplies of *Hevea* rubber from Southeast Asia were cut off, the U. S. government established a guayule rubber project in which more than 32,000 acres of guayule were planted. This project produced over a billion seedlings and 3 million pounds of rubber for the war effort. This project was abandoned after the war when rubber from the *Hevea* tree became readily available. Because of the changing economic situation and the uncertainty of continued availablility of raw materials for synthetic rubber, guayule research was renewed in the late 1970's. Most important from this research is the discovery by U. S. Department of Agriculture investigators that certain low-cost triethylamines, particularly 2-(3,4-dichlorophenoxy) triethylamine [825, 826], sprayed on the plants about three weeks before harvest increased yields two to six times the normal 500 lbs. per acre. Estimates indicate that the process of using such chemicals could increase overall yields 30 to 35 percent and cut the growing time by one to two years [827]. These results came from the study by this group, led by chemist H. Yokoyama, of the regulation of biosynthesis of carotenoid pigments in fruits and vegetables. Most of the carotenoids are of the *trans* configuration. Those with the *cis* configuration are known to exist in nature, but they are rare. One example is *cis*-lycopene, which occurs in the tangerine tomato and has a light orange color [828].

Yokoyama and his colleagues discovered a number of compounds that de-repress the synthesis of enzymes involved in the biosynthesis of *trans*-carotenoids in citrus fruits; for example, when a lemon is treated with such a compound, the synthesis of *trans*-lycopene is enhanced and the fruit develops a deep red color. He also discovered a compound that de-represses synthesis of enzymes involved in the biosynthesis of *cis*-carotenoids. When he sprayed this on the lemon, the lemon turned orange due to predominance of *cis*-lycopene.

Carotenoids and natural rubber both belong to the same family of products derived from isoprene. Natural rubber is a *cis*-polymer of isoprene. When Yokoyama became aware of the revived interest in guayule, he arranged to have a number of young guayule plants to work with. When he applied the *trans* activator, he found it to be inactive. In other words, he found those which were inactive on citrus fruits were also inactive on guayule. The *cis* activator, however, increased the rubber content two- to threefold. These results suggest that rubber productivity can be improved on other rubberforming plants such as *Euphorbia* as well as the *Hevea* and guayule plants [825].

Pine – Oleoresin Accumulation

In the early 1970's, it was reported that the application of paraquat to the exposed wood of several species of pine caused the accumulation of oleoresin in the boll of the trees [829]. Paraquat enters the transpiration stream and causes the living xylem cells to synthesize large quantities of oleoresin that are subsequently transferred (secreted or leaked) in large amounts to neighboring cells until the entire area is saturated [830–833]. This accelerated process of "lightwood" formation has stimulated considerable interest in the industrial community, resulting in a substantial amount of applied and basic research on the use of paraquat as a stimulator of oleoresin in pine [834–839]. Again, the interest here is as an enhancer of oleoresin content in pine – providing the possibility of an alternate source of chemical feedstocks for industrial usage to supplement those obtained from crude oil. The attractiveness is because the hydrocarbon supplied by the oleoresin represents a renewable resource. This potential also serves as a boost to the naval stores industry, which has been on the decline, as it holds the promise of a return to production levels experienced during the first half of this century [840].

Apples

There have probably been more investigations with the plant growth regulators on apples than on any other crop – and with considerable success, as can be seen throughout this book. However, it is only recently that efforts have been made specifically to improve fruit quality, although pomologists have long recognized that plant growth regulators, when used for a variety of purposes, also affect fruit quality both at harvest and postharvest.

The introduction of daminozide in the mid 1960's was an important contribution to this area of research [841]. When applied up to mid-summer, daminozide improves the color of red apples at harvest [842, 843]. Although it delays development of such quality factors as aroma and juiciness, sugar and acid levels are not affected. Daminozide has been found to reduce the development of water core [844] and superficial scald [843] on "Delicious" apples. It has been found also to increase firmness of fruit both at harvest and after periods of storage in the cold [842, 844, 845].

The subsequent availability of ethephon in the latter part of the 1960's also had major effects both in the control of fruit ripening and in improvement of fruit quality in apples. Ethephon dramatically improves the red coloration of certain apple cultivars and rapidly transforms green fruit to ripe fruit for early marketing [846–852]. Since ethephon-treated apples are apparently indistinguishable from untreated fruit harvested several weeks later, the growers are able to extend their marketing season as a result of ehtephon treatment. Where color is a consideration, growers benefit by the ethephon effect on coloration.

Sometimes the use of ethephon results in a softening of fruit and in abscission. This can be countered by the previous application of daminozide to such trees, particularly in the case of "McIntosh" [847, 849, 851, 852].

A recent addition to the growth-regulator treatment of apples is a combination product of benzyladenine plus a mixture of gibberellins A4 and A7. This combination proprietary product is marketed as Promalin by Abbott Laboratories. Applied at full bloom the shape of "Delicious" apples is changed by increasing the ratio of length to diameter, resulting in an improvement in this desirable quality [853–856].

Apricots

It has been known since the 1950's that both 2,4,5-T and 2,4,5-TP increase the size of apricot fruits and advance their maturation [857, 858]. The use of 2,4,5-TP has become standard practice in apricot orchards of Israel. There are numerous reports that daminozide accelerates the ripening of stone fruits [859–862], but these experiments were primarily with peaches and cherries and not with apricots. Recent reports from Israel with daminozide applied to apricots show an enhancement of color when applied at or after pit hardening [863]. Surprisingly, however, the other aspects of ripening were either unaffected or even retarded, suggesting that daminozide has different effects on different species of the stone fruits.

Bananas

It has been known for almost forty years that 2,4-D can be used to aid the ripening of bananas [864–866]. More recent work has shown that the application of 2,4-D improves the color of fruit, producing a deeper yellow color regardless of the season applied [867]. Such treatments are usually equally effective on fruit throughout the bunch, thus resulting in more uniform ripening; however, the treatment with 2,4-D results in an acceleration of ripening by about four days and this should be kept in mind from a commercial point of view.

For decades it was known that low levels of ethylene are sufficient to initiate prompt ripening of bananas [868–871]. Thus, it is hardly a surprise to find that ethephon induces ripening in bananas, producing a superior grade of fruit compared to those from normal ripening [872–874].

Recent studies with a number of plant growth regulators show that abscisic acid and indoleacetic acid hasten the ripening of bananas, whereas dip treatment with gibberellic acid and kinetin retarded the ripening [875].

Berries, Cane

Several studies have been conducted to determine the effect of daminozide on red raspberries with variable results – some increasing yield [876] and others having very little effect [877, 878]. Preharvest applications of ethephon increased the rate of maturation of red raspberries [879]. Similar results have been reported for blackberries [880]. A recent study shows daminozide, when applied both at full bloom and three weeks after bloom, to delay ripening of blackberries [881]. Ethephon and dikegulac had little effect in these studies.

Blueberries

The harvesting of highbush blueberries in Nova Scotia is manual and is largely dependent upon school children. Because school schedules conflict with the harvesting schedules, studies were conducted with ethephon and results showed that ethephon reduced the fruit removal force and allowed machine harvest of the majority of the crop [882], confirming previous results from Michigan [883]. Early maturing varieties are of little help because the principal market is after the opening of school. Hence, the market is best served by late-season cultivars such as the "Coville". Eck [884] showed that the application of ethephon to highbush blueberries hastened fruit ripening and increased percentage of ripe fruit at the first harvest, but produced smaller berries. Recent work shows that treating "Coville" blueberries with ethephon to reduce the fruit removal force does not have an adverse effect on either quality or yield [882].

Rabbiteye blueberries are native to the southeastern United States and have recently received attention because they represent a cash income in developing rural areas in that region. Large variability in the maturation of the fruit occurs among the several varieties. If the blueberry could be made to ripen most of its fruit for a single harvest, an increase in the effectiveness of mechanical harvesters would be realized. Work in Georgia [885] shows that application of daminozide and ethephon as preharvest sprays resulted in fruits that ripened significantly earlier than untreated ones. Multiple applications of ethephon as well as the application of daminozide followed by ethephon brought 95 percent of the fruits to full ripeness, reducing the length of the harvest period by approximately a week and with no significant effect on the size of the berries. Fruit coloration was increased and storage quality was maintained for forty days, a prolongation of the marketable life. Recent results from Georgia indicate that treating of certain rabbiteye blueberry cultivars with gibberellic acid produced fruit with fewer seeds without reducing the total yield and apparently not affecting the shelf-life of the blueberries [886].

Cherries

Daminozide applied shortly after bloom advances anthocyanin pigmentation in sweet cherries by about two weeks [848, 887]. Sugar levels comparable to fully mature fruit are attained in treated fruit one week earlier than normal. Because the respiration rate of sweet cherries is not affected, it is assumed that the applied daminozide affects carbohydrate metabolism, encouraging an increase in sugar and the development of anthocyanin. Daminozide advances the color, the size, and soluble solids in sweet cherries without altering the fruit acid levels [848]. When applied prior to harvest, daminozide delays maturity [888].

The reverse effect is obtained with gibberellins. The application of gibberellic acid to sweet cherries delays harvest up to five days. The treated fruits are firmer and withstand handling and storage better than untreated ones. In addition, by remaining on the tree a few days longer, the fruit tend to be larger [848].

As with sweet cherries, daminozide advances the maturity of sour cherries [848]. This includes reducing the fruit removal force, increasing color and size, and advancing harvest date by at least one week. An example of this effect is shown in a Michigan orchard where the improved fruit condition due to treatment with daminozide resulted in the harvest period for a single orchard being safely spread over a four-week period as contrasted to the normal two-week period.

Results similar to those produced by daminozide can be obtained with ethephon applied two to three weeks before harvest [889, 890]. Specific effects include reduced fruit removal force, more uniform ripening, increased and earlier coloring, and overall more rapid maturation. Because its most notable effect is on abscission, ethephon also is useful as an aid to mechanical harvesting of cherries [848].

Citrus

The navel orange fresh fruit market is an important segment of the California citrus industry. For optimal marketing purposes, it is necessary to store the crop on the tree and harvest it over a period of from four to six months after the crop reaches legal maturity. Although there are a few problems early in the harvest season with the rind, a greater number of disorders appear during the latter part of the harvest season when the fruit is still of prime eating quality. Most of these late-season physiological disorders appear to be a result of aging of the rind. One such problem, and a serious one, is called "rind staining". It has been shown that preharvest application of gibberellins causes a significant reduction in rind staining of navel oranges [891–893]; when applied to green fruits, there is a delay in the loss of chlorophyll pigments [894]. Another disorder that develops with age is a stickiness of the rind. This condition is markedly reduced by preharvest application of gibberellin [895]. The correction of this disorder is important because firmness to the touch affects the ultimate marketability of fruit.

Creasing is a disorder of "Valencia" oranges, due to a puffiness of the peel, causing marked economic loss in Israel. It was found that gibberellins increase the viability of the peel by rendering the tissues more compact and by delaying senescence [896, 897]. This is done without hampering the color break or carotenoid accumulation.

Some fruits of the "Shamouti" orange in Israel do not meet export standards when produced under conditions that cause them to be excessively large, thick peeled, and rough; this is especially so when the crop is grown in marginal soil and in an arid climate. This condition can be overcome by early sprays of daminozide and chlormequat [896, 898, 899]. It is postulated that these plant growth regulators are effective by counteracting the high endogenous level of growth promoters found in the rough tissues of the orange peel.

Coffee

Although the optimal concentrations may vary from area to area, ethephon can be used to effectively ripen coffee berries [900–903]. During the period in which ripening is enhanced by the application of ethephon, harvesting efficiency should also increase. Positive results of this nature have been reported from Rhodesia, Puerto Rico, and Hawaii.

Corn

Probably the greatest publicity ever received by a plant growth regulator on a food crop was that of dinoseb and its effect on corn. Many reports both in the lay and scientific literature extol the virtues of this compound and its effects in increasing the yields of corn [904–908]. The first observations of the stimulatory effect of dinoseb on corn were made in 1968 in field experiments at Purdue University and reported by Hatley in his thesis in 1970 [909]. This stimulation resulted from incorporation of dinoseb in the fertilizer band. All subsequent work has been with foliar applications. Since the original report of the effect of dinoseb, investigators in various parts of the world have published conflicting reports on the effectiveness of the compound [910–917], possibly because of variations in genetic background of the different corn varieties used. Nevertheless, three American companies (Dow Chemical, Helena Chemical, and Agway) are marketing formulations of this chemical for use on corn.

Dinoseb was used by a number of U. S. corn producers between 1975 and 1980, but its use is declining. In his summary of the use of plant growth regulators in corn at the 1980 meeting of the Plant Growth Regulator Working Group, Oplinger suggested that the success of dinoseb "has been limited in part because it lacked the backing of a major chemical manufacturer and also because it does not appear to be an effective yield enhancer on all corn hybrids" [918].

A recent report from Czechoslovakia states that the addition of carbofuran to the seed dressing of corn stimulated germination and subsequent growth of corn and increased the yield of both silage and grain [919].

Cucurbits

Lukasik [920] in Poland conducted an extensive study with a number of plant growth regulators sprayed on seedling cucumbers to determine their effect on yield. The highest yield obtained in experiments conducted over several years was from plants treated with 2,4,5-trichlorophenoxy propionic acid. The next best results were from those seedlings treated with 2,4,5-trichlorophenoxyacetic acid and naphthaleneacetic acid. In years with generally cold temperatures, gibberellic acid had a positive effect. On the other hand, in warm years treatment with gibberellic acid gave the lowest yield in the experiment.

In muskmelons, a primary criterion for the evaluation of fruit quality is the soluble-solids content of the melon at harvest. Numerous chemical substances have been tested in attempts to increase the soluble-solid contents of this melon without much success. Because of the positive effects with glyphosine in increasing the soluble-solid contents of sugarcane, particularly during unfavorable ripening conditions, this compound was evaluated for its effect on muskmelons. Soluble solids were found to be increased at all concentrations used when glyphosine was sprayed on muskmelon after initial flowering [921]. The branch length and number of leaves were reduced at the higher concentrations (1600 ppm); the melon weight was increased at the lower concentrations (200 ppm). Both of these effects were more evident toward the end of the season. In general, at most glyphosine concentrations, the soluble-solids content was increased about 10 percent above untreated controls. In the same experiment, treating muskmelons with various concentrations of triacontanol as foliar sprays produced no effect.

Grapes

One of the more outstanding successes with plant growth regulators has been the use of gibberellin on "Thompson Seedless" grapes. Within a few years after the first experiments with gibberellin on this variety [922–924], practically all "Thompson Seedless" grapes intended for table use were being sprayed with gibberellin at the fruit set stage to increase the size of the berries. Commercially, gibberellin is now applied two times – the first one at anthesis, the second at the fruit set stage [925]. There are other positive effects obtained by these sprays in addition to an increase in berry size.

A group of Australian workers were the first to show that ethephon hastens maturity of grapes [926] and this was quickly confirmed by California investigators [927]. Ethephon enhances the coloring of "Emperor", "Tokay", and "Red Malaga" grapes [927–930]. Total acidity is not affected in the former two varieties although it is decreased in "Red Malaga" [930]. Recent work in India with purple grape varieties show that ethephon sprays about four weeks before harvest effectively reduce the percentage of unripe berries and improve the quality of the harvested grapes [931, 932].

One of the classic examples of the control of fruit set and development and the advancement of maturity is the use of gibberellins for the production of seedless "Delaware" grapes in Japan [933]. "Delaware" is the most important table grape in Japan, and the use of gibberellin to induce seedlessness is a standard commercial practice in that country. When properly applied, essentially 100 percent seedlessness can be obtained. Such results have been confirmed in the United States [934]. Interestingly enough, the use of gibberellin to produce seedlessness in other varieties of grapes has not proved to be commercially successful.

Recent work has shown that spraying with solutions of vanillin or ammonium isobutyrate, both of which have been demonstrated to be ripening agents for sugarcane, increases the sugar content of grapes [935].

Nuts

Work in Iran shows that spray treatment of pistachio nut trees one month before normal harvesting greatly accelerates the maturity of the nuts [936].

Peaches

The harvest maturity of peach fruit can be hastened by the application of 2,4,5-T, but undesirable fruit quality often results [937, 938]. Daminozide was been reported to accelerate peach ripening without adverse effects [939–944]. In addition to hastening fruit ripening, daminozide increases internal flesh color and skin color.

Ethephon causes an increase in peach fruit size [945, 946] in addition to accelerating fruit maturity [941, 947, 948].

Either compound can advance the commercial harvest and promote the ripening of peach fruit throughout the final growth period, effects reversed by gibberellic acid [949]. The two in combination or in successive applications are more effective in promoting early maturity than either chemical alone [944].

Field spraying of immature peach fruit, with daminozide, ethephon, or gibberellic acid leads to a marked reduction in browning of the fruit when pureed or sliced after harvest [950]. This failure to darken was shown to be due to a 90 percent reduction in the activity of the enzyme polyphenoloxidase in the mature fruit [951].

Peanuts

Although most investigators agree that plant growth regulators show considerable promise for use in peanuts, the results obtained to date are variable. Das Gupta in Sierra Leone [952] found that foliar sprays of chlormequat at the proper concentration gave more kernels and heavier yields of pods, kernels, and oil. Wittwer [412], in his summary of the uses of daminozide, reported that daminozide on peanuts gave rise to increased yields, higher grade nuts, and plants that showed greater drought resistance. Brown and his coworkers [953, 954] summarized the overall effects of daminozide on peanut plants as: (a) reduction in plant size, (b) darker color and thicker leaflets, and (c) shorter and thicker pegs with pods closer to the central branches. In this same study they found yield increases due to an increase in number of pods per plant. Other authors [955–958] have found either no effect or slight increases in yield from the use of daminozide, although grades of daminozide-treated peanuts tended to be lower. Gorbet and Rhoads [959] reported that treatment with daminozide reduced the water required to maintain the desired water level in the soil of irrigated plots of peanuts. Wu and Santelmann [958] tested several plant growth regulators for their effects on peanuts. These tests included daminozide, mefluidide, TIBA, ethyl-5(4-chlorophenyl)-2H-tetrazol-2-yl-acetate, 2-chloro-3-(3-chloro-2-methylphenyl)propionitrile, as well as compounds whose structures were not available. None of the treatments increased the yield of peanuts in any of three consecutive years. These investigators felt that, although no significant yield increases were obtained, the reduction in vine growth as a result of treatment might be useful in areas of high rainfall.

TIBA application to peanut plants caused changes in vegetative growth and variable effects on yields [955, 960–962]. Morphactins stimulated pegging, altered apical dominance, promoted lateral branching, and reduced peanut yields [962, 963].

Working with "Starr" Spanish type peanuts, Ketring [962] found when plant growth regulators were applied at early flowering, shoot fresh weight at harvest was reduced by a morphactin but was increased by abscisic acid and picloram + 2,4,5-T. None of these compounds affected growth when treated at late flowering. He also found no apparent relationship between vegetative growth and weight of mature seeds produced. In all treatments, the seeds produced either germinated abnormally or had reduced vigor.

Pears

The control of natural ripening in pears is thought to be accomplished through endogenous growth regulator mechanisms, with ethylene being the predominant compound involved [964]. Hansen [965, 966] showed that ripening of pear fruits is induced by both auxin and ethylene and that endogenous ethylene production is essential for proper ripening. Looney's work [967] shows that daminozide delays the ripening of pears by suppressing ethylene levels.

Ripening inhibitors are essential to promote the shelf-life of pears and to control problems such as premature ripening. Gibberellic acid and daminozide counteract the accelerated rate of pear ripening [967, 968]. Other reported inhibitors of ripening in pears include cycloheximide [969], indoleacetic acid, 2,4-D, alpha(p-chlorophenoxy)isobutyric acid [970, 971], and aminoethoxyvinylglycine [972, 973].

Peppers

Early work showed an increase in red color of pimiento peppers through the use of ethephon [974]. Subsequent work in Michigan with paprika peppers [975] showed that ethephon stimulated ripening of the fruit and increased the total extractable color. Because paprika for food coloring is sold by unit weight and the market value is determined by pigment content, increase in the amount of extractable color increases the market value of these peppers. Similar work in Argentina with red peppers [976] showed ethephon treatment as a spray to increase significantly the percentage of mature fruit at harvest.

Recent results with banana peppers in Arkansas [977] show only negative effects on yield from ethephon applied either as a soil treatment and as a foliar spray. However, since the negative effects decreased as the concentration was decreased, the investigators feel that the proper level had not been reached and might be lower than those used in their studies.

Plums

Early-ripening plums of the Italian prune type are grown in the Pacific Northwest United States mainly for fresh use or for canning rather than for drying [978]. These varieties are subject to internal browning, a disorder that appears in storage [979]. Treatment of trees of these varieties with sprays of gibberellic acid before harvest result in firmer fruit with less internal browning and longer shelf life [980].

Studies in California with Japanese plums [981] show that a preharvest application of ethephon hastens skin color changes and flesh softening but does not hasten soluble solids accumulation. Post-harvest evaluations indicated similar ripening rates for both treated and untreated fruit.

In Australia, plum trees treated with daminozide and ethephon [982, 983] bore fruit slightly smaller than the controls, whereas those treated with gibberellic acid and naphthaleneacetic acid were found to have fruit slightly larger than the controls. Although ethephon and daminozide advanced fruit maturity of plums, they are not recommended because they cause too many undesirable features. Gibberellic acid and naphthaleneacetic acid are recommended for use on plums; the delay in harvest time and the increase in size are stated to be financially worthwhile.

Soybeans

High yields of sexual fruits and vegetables depend on the ability of the plant to convert energy normally used for vegetative growth into seed. The efficiency with which this is done is a major factor in determining yield. This is specially true of soybeans, which produce more vegetative growth than can normally be used for seed production. A growth regulator applied to suppress excessive vegetative growth should allow more energy to be used for reproductive processes and more flowers to be formed, creating the potential for increased yields.

The earliest work was probably that of Van Schaik and Probst [984] who studied six growth regulators but found none to be effective in increasing pod set in soybean.

In the 1960's, TIBA was considered a spectacular plant growth regulator. When applied to the leaves of soybeans, it shortens the plants, increases their branching, stiffens them, and increases pod set, thus increasing yield [985]. Unfortunately, the compound is not always effective. Scientists have found that the timing of its application is critical and that different varieties respond differently. Hence, results are disappointingly inconsistent and the use of TIBA to increase soybean yields has been discontinued even though it is registered for this use.

Because of the importance of this crop, studies have continued evaluating plant growth regulators for their potential in increasing soybean yields. Early in this search it was recognized that different plant types would respond differently to different types of chemicals and that differing application rates might be required for maximum plant response in a given set of environmental conditions. Thus the importance of plant types (determinant or indeterminant) was realized as was their physiological interactions with a changing environment. A chemical that might be effective on a determinant type of soybean might not be effective on an indeterminant type, and vice versa. An ex-tensive program for screening and evaluating plant growth regulators for their potential effects on soybean has been established by Stutte and co-workers at the University of Arkansas [986–990]. The most active materials found in their studies were tributyl(5-chloro-2-thienyl)phosphonium chloride, tetrahydrofurfuryl isothiocyanate, and sodium-1(p-chlorophenyl)1,2-dihydro-4,6-dimethyl-2-oxonicotinate.

Interestingly enough, they found that a number of plant growth regulators have no effect on yield of soybeans. These included gibberellic acid, TIBA, ethephon, 2,4-D, benzyladenine, maleic hydrazide, chlormequat, and daminozide.

Tobacco

The growth regulator ethephon was initially registered and recommended for tobacco in southeastern United States to reduce the number of primings and the time required to cure the crop. Since the work of Steffens and his colleagues in 1970 [697], most workers agree that the efficacy of ethephon as a field ripening agent on tobacco leads to fewer primings and quicker curing [991–996]. The latest confirmation of these results is from India [997].

Tomatoes

Ethephon [998–1004], the methyl ester of chlorflurenol [1005], 2-chloro-ethanethiophosphonic acid dichloride [1006], and 2-amino-6-methylbenzoic acid [443] have all been shown to accelerate the ripening of tomatoes and to increase the yield of marketable fruit.

Miscellaneous Vegetables

In 1963 Supriewska [1007] showed that chlormequat increased the dry weight of storage roots of carrots. This effect of chlormequat on carrots was confirmed and similar effects were shown with daminozide [1008, 1009] and ancymidol [1010]. The most recent work from the National Vegetable Research Station in England shows three growth retardants to be effective in increasing the root growth of radish, with the most effective being daminozide [1011], confirming earlier work in Idaho [1012].

Studies in Egypt with cauliflower showed that foliar sprays of NAA increase the total yield of cauliflower plants as well as improves diameter, weight and color of the curd. The best curd compactness was obtained by treatment with IBA [1013].

Anti-Ripening Agents

While most of the discussion in this chapter has been concerned with the use of plant growth regulators for their effect in the control of maturation or ripening, sometimes these compounds can be used to negate the ripening effect of another plant growth regulator or the natural ripening of a fruit or vegetable. For example, the use of auxins such as NAA, 2,4-D, and 2,4,5-T on apple trees to check the preharvest abscission of fruit causes a stimulation of respiratory activity in certain varieties. This stimulation results in shortened storage life for the apples and, of course, detracts from the commercial use of such materials. As early as 1947, Smock and his co-workers [1014, 1015] showed that spraying maleic hydrazide along with one of the auxins on apple trees minimized the ripening effect while the desired effect of the auxin was not interfered with.

Recently, it has been shown that certain dialkylaminoethylamides are useful in retarding the ripening of picked fruits and vegetables [1016–1018].

Fig. 17-1. Fixed-wing biplane applying ripening agent to sugarcane field in Hawaii

Fig. 17-2. Sugarcane stalks taken to analyze for effects of ripener

Fig. 17-3. Effect of ethephon on internodal elongation in sugarcane, increasing bio-mass. Photograph courtesy of Union Carbide Agricultural Products Company

Fig. 17–4. Hevea rubber tree showing tapping panel (where ethephon formulation is painted directly on the tapping cut) and the latex collecting cup. Photograph courtesy of Union Carbide Agricultural Products Company

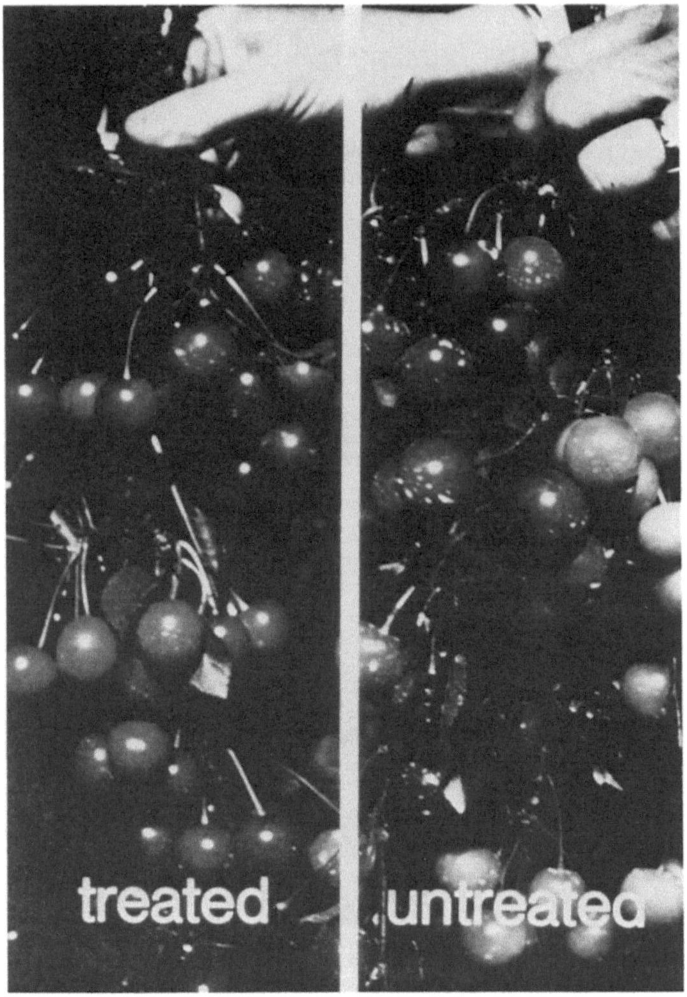

Fig. 17-5. Uniformity of ripening in cherries by treatment with daminozide. Photograph courtesy of Uniroyal Chemical

Fig. 17-7. Ripening effect of ethephon on tomatoes showing reduction in discarded green tomatoes following treatment. Treated tomatoes at right; untreated at left. Photograph courtesy of Union Carbide Agricultural Products Company ►

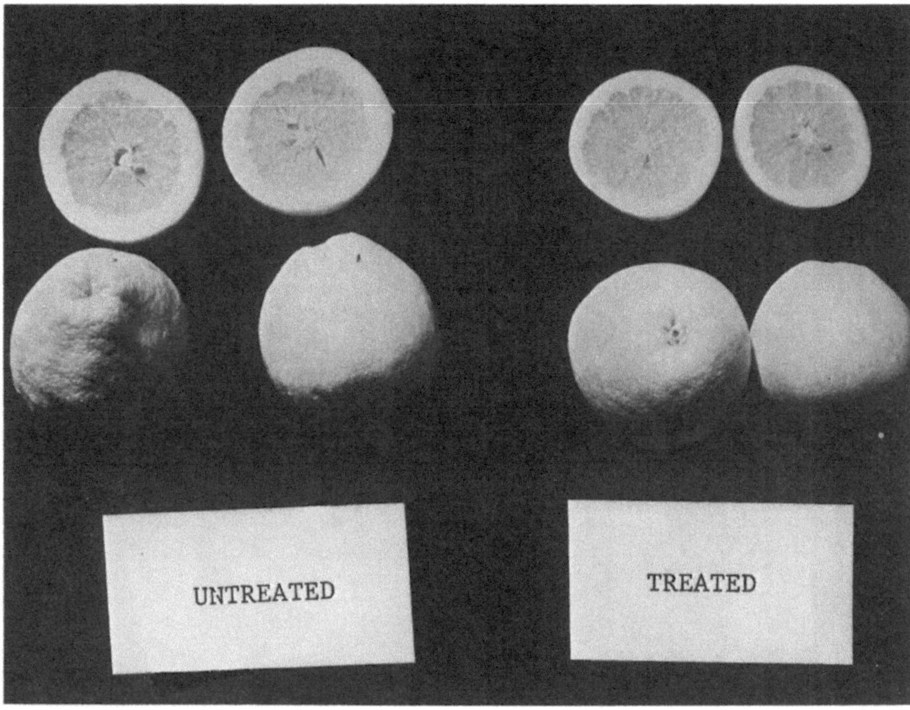

Fig. 17-6. Effect of gibberellic acid on grapefruit shape, rind quality, and internal structure. Photograph courtesy of Abbot Laboratories

Fig. 17-8. Effect of 2-(3-Chlorophenoxy)propionic acid on pineapple, showing delay in maturity, inhibition of crown, and increased fruit size. Treated fruit at left; untreated at right. Photograph courtesy of Union Carbide Agricultural Products Company

Chapter 18. Modification of Sexual Expression

Being able to manipulate the sexual expression in plants has practical implications both for facilitating seed production and for increasing yield. For example, when the yield of a plant is limited by the number of female flowers it produces, any shift to increase the femaleness should result in an increased yield.

Plant growth regulators induce flowering in certain plants, prevent flowering in others, and alter the number of flowers; in addition, these regulators can also alter the sex. Laibach and Kribben [1019, 1020] reported that treating gherkins with auxin increased the number of female flowers and reduced the number of male flowers. Work by other investigators showed that applying auxins shifts the balance of sex expression from maleness to femaleness in certain plants [1021–1024]. Subsequently it was found that applying gibberellins induces the formation of male flowers in certain flowering plants [1023–1031]. Although not equally effective for all varieties of cucumbers, ethephon has been shown to promote female flowers in many members of the cucurbit family including cucumber [1032–1042], pumpkin [1043, 1044], squash [1033, 1034, 1040, 1045, 1046], gourds [1047], and various melons [1048–1053].

In addition, although emphasis has been on the sex effect in cucurbits, ethephon has been shown to induce female flowers in male plants of *Cannabis sativa* [1054], whereas gibberellin induces male flowers on female plants [1055]. The implications of this ability to favor one or the other of the sexes is important as a tool for hybridizers in breeding and as a means of increasing production. Corley [1056] found that gibberellic acid increases the production of male inflorescences in the oil palm and reduces the production of female inflorescences; he also found that ethephon and NAA also have effects on sex ratio, but concluded that these are indirect effects.

Treatments with growth retardants that interfere with gibberellin biosynthesis tend to induce a shift towards femaleness in cucurbits [1057, 1058]. Investigators in India [1059] have found that gibberellic acid increases the proportion of male flowers in coriander whereas ethephon and chlormequat decrease both the total number of flowers and the proportion of male flowers.

Morphactins also cause a shift towards maleness in flowers of cucurbits [1047, 1060]. Early work with triiodobenzoic acid (TIBA), which inhibits auxin transport [1061] was reported to increase the femaleness of two cucumber cultivars [1062]. Later work showed that TIBA causes the production of predominantly male flowers on monoecious cucumber plants treated at the first true leaf stage [1038]. In the same study ethephon caused predominantly pistillate flowers on treated plants.

Work with excised embryos of hemp shows that gibberellic acid treatment resulted in male mature plants whereas benzylaminopurine resulted in female mature plants [1063]. In this study gibberellic acid also hastened the onset of flowering. The application of benzylaminopurine to male inflorescences of gourd induced bisexual and female flowers [1064].

Studies in India with the tropical fruit, ber (*Zizyphus mauritiana*) [1065] showed that the cumulative effect of chlormequat and naphthaleneacetic acid induced early and profuse flowering with an increase in female flowers.

Commercial hybrids of muskmelon have gained wide acceptance, but seed is expensive because of the necessity for hand pollination. Recent studies suggest the feasibility of using gynomonoecious parents for producing hybrids without hand pollination [1053].

These studies showed that ethephon applications extended the pistillate flowering stage in these lines of muskmelon. Foliar application of aminoethoxyvinylglycine enhanced staminate and/or perfect flower formation in these lines [1066].

5-Methyl-7-chloro-4-ethoxy-carbonylmethyl-2,1,3-benzothiadiazole (TH-6241) was found to promote maleness in both cucumber and muskmelon [1067].

Aminoethoxyvinylglycine, an inhibitor of ethylene production, increases the number of staminate flowers in gynoecious plants treated with this compound [1066, 1068, 1069], as does a mixture of gibberellins A4 and A7 [1028, 1033, 1068, 1070]. Silver nitrate which has been shown to inhibit the action of ethylene also causes this effect.

The current theory is that the auxin/gibberellin balance in cucurbits is the physiological basis for their sex determination. Studies at Michigan State University [1067] suggest that the endogenous concentration of auxin may in fact determine the endogenous concentration of ethylene. Ethylene, then, would be an intermediate in the promotion of femaleness.

Studies at the New York Agricultural Experiment Station conclude that in cucumbers changes in both sex expression and fruit shape are probably controlled by the same endogenous hormone [1042].

Chapter 19. Senescence

The potential for plant growth regulators to regulate senescence is becoming apparent. Investigators in this field feel that the senescence signal is a hormone-like substance that travels from the fruit to the leaves and other vegetative parts where it triggers senescence [1071]. It is well known that abscisic acid prompts senescence [1070, 1072] and that a wide range of species is sensitive to this material [1073]. An extensive literature has developed identifying cytokinins as the most generally effective class of senescence-retarding growth regulators [1074–1077]. For example, NAA used together with benzyladenine effectively delays senescence in soybeans [1078]. Applications of auxins or gibberellins have been reported to delay senescence in certain species such as oranges [1079–1084]. Rhizobitoxine analogs have been shown to extend the shelf-life of broccoli [1085] and the vase-life of cut flowers such as carnations [1086], iris [1087], daffodils [1087], chrysanthemums [1087], and snapdragons [1088]. Daminozide, when included in the sucrose-holding solution, markedly extended the longevity of carnations and extended the longevity of roses by three days. Cycloheximide was effective in increasing the longevity of both flowers but tended to be phytotoxic [1089].

The ability to supply domestic and export markets with attractive, high-quality fruit over a relatively long period of time is important to the success of any fresh fruit crop. The California citrus industry enjoys considerable success in being able to place high-quality fresh fruit on the market during a major portion of the year. Part of this success is due to the use of growth regulators to prolong the pre-harvest and post-harvest life of lemon [1083, 1090], lime [1090], grapefruit [1083], and orange [1079–1083] fruits. The summary by Coggins [1083] covers this very well: "Separate and combination pre-harvest applications of gibberellic acid and 2,4-D are used widely in California to prolong the pre-harvest and post-harvest life of navel orange fruit. 2,4-D is used to delay fruit abscision and gibberellic acid to delay fruit maturation, preharvest applications of 2,4-D to reduce drop of mature lemons, and post-harvest 2,4-D treatments to delay senescence of the fruit button and thereby reduce decay during storage. Although 2,4-D is used widely in grapefruit and "Valencia" orange crops to reduce drop of mature fruit, gibberellic acid is not in general use on these crops due to an increase in regreening of rind tissue. The delay in rind senescence, however, is of value to these crops." Gibberellic acid is widely used by California navel orange growers to improve fruit quality by delaying rind senescence and thus reduce rind staining, water spot, susceptibility to decay, and development of sticky rind surfaces [1090–1092].

The winter crop of lemons in Australia matures in June and if not pikked by early September, a large portion of it will drop after becoming tree-ripe. The remainder of the fruit will substantially increase in size, rendering it unacceptable as fresh fruit. Studies with gibberellic acid and chlormequat showed that a combination of these two growth regulators delayed coloring and arrested the growth of lemons sufficiently to extend the harvest into the late spring by increasing the supply of small-sized premium quality fruit [1093]. There are no regulations in Australia for color and size of fresh lemons, nevertheless, premium prices are paid for fruit of proper size and color.

Chapter 20. Desiccation

There are several reasons for desiccating specific crops prior to harvest. For crops where the economic crop is the seed, such as the legumes (particularly soybeans), desiccation hastens maturity so that seeds can be harvested earlier than usual. The same is true for sunflower with the additional advantage of simultaneous weed desiccation. Growers who "double crop" find more time to prepare and plant the second crop, such as winter wheat following soybeans, if they induce faster bean dry-down for earlier harvest. The use of desiccants has significantly increased the "double crop" acreage. In cotton, the aim of desiccation is to remove the leaves for efficient mechanical harvesting of the cotton bolls; in potatoes, it is to eliminate as much of the above-ground portion of the crop as possible prior to harvest. In the case of rice grown in the United States, preharvest desiccation would improve the timing of harvest and allow more efficient and faster combining of the grain. This would allow more grain to be harvested in a given time period, not only increasing the efficiency of the harvesting equipment but allowing the crop to be removed prior to the advent of tropical storms that reduce both yield and quality because of lodging and weathering.

Cotton

In cotton, there is an overlap between defoliants and desiccants, with some compounds producing both effects or overlapping effects. In some areas of the cotton-growing regions in the United States desiccation is a substitute for defoliation and is used for rapid drying of the leaves and other plant parts prior to harvest. This practice is standard in the high plains in Texas, Oklahoma, and New Mexico where stripper-harvest methods are used [235]. For this method of harvesting, the complete removal of, or the drying of, leaves is essential for satisfactory results. Because defoliants often do not cause complete leaf removal, desiccants are used extensively to allow harvesting before frost. Desiccants are also used in other cotton growing areas of the United States if conditioning is needed late in the season when defoliants act too slowly or when there is considerable regrowth. Another advantage of desiccants in cotton is that they can be used effectively nearer to scheduled harvest dates than can defoliants [1094], thus allowing additional time for maturation of late bolls [1095]. Recent results have shown that desiccation accelerates the drying process in the green bolls and hastens their opening with no deleterious effect on quality if the bolls are otherwise mature [1096]. Similar results have been reported in India, where the two chemicals used primarily for desiccation are paraquat and arsenic acid [1097] (either of these compounds can also cause leaf abscission).

Soybeans

The use of desiccants in soybeans has several advantages. Harvesting can start as much as six days earlier than usual; this earlier harvest can have a marked effect on crop quality, yield, and the ultimate price received for the soybeans. Early harvest can minimize crop loss due to shatter and can help prevent infestation by fungi [1098]. Several chemicals have been successfully used for soybean crop desiccation: ametryn, glyphosate, and paraquat [1099–1101]. Ametryn is currently registered for use as a potato desiccant. Glyphosate shows promise for hastening maturity in grain sorghum. Paraquat is the most widely used desiccant for soybeans, having been registered for this use for over ten years [1101].

Rice

The value of a preharvest desiccant for rice has been recognized for many years. In addition to the desirability of decreasing the amount of green leafy material at harvest as would be the case with most other crops, there is the added need to shorten the harvest time to complete the process before the occurrence of tropical storms in coastal areas of the United States. Many coumpounds have been evaluated as preharvest desiccants on rice including: DEF [1102], dinoseb [1103], endothall [1102, 1104], magnesium chlorate [1102], sodium chlorate [1104–1106], sodium chloracetate [1102–1105, 1107], sodium pentachlorophenate [1103], and trichloroacetic acid [1103]. Each of these desiccants has one or more disadvantages including reduced yield, abnormal color of grain, off-flavor grain, and residues in the grain. Recent work with diquat, paraquat and sodium chlorate showed them to be preferable as rice desiccants [1108, 1109]; at present, however, only sodium chlorate is registered for this use. Studies in India with both diquat and paraquat in paddy rice show effective desiccation at appropriate levels with residues found that are within the tolerance limits prescribed by WHO/FAO [1110].

Potato

In the United Kingdom, potato vines are burned off using sulphuric acid, diquat, or dinoseb. These are quick-acting contact herbicides with desiccant properties but, under unsettled weather conditions, retreatment is sometimes necessary. On the other hand, metoxuron is slow acting and can be added with the spray for the late blight fungus. In Poland, ethephon showed a consistent and satisfactory desiccation of potato vines; these studies were compared with diquat-treated plants, and both compounds were found to be equally effective [1111]. In the United States the most effective compounds are paraquat [1101] and 2,3-dihydro-5,6-dimethyl-1,4-dithiin-1,1,4,4-tetraoxide [1112–1114]. A recent report from Germany shows 2,5-dichloro-4-nitrophenol to be an effective desiccant for potato vines [1115].

Grain Sorghum

For many years the need has been recognized for a safe, effective chemical for the preharvest desiccant of grain sorghum. This would promote earlier ripening and reduce losses both to weather and to pests. Although diquat and paraquat were found to be effective on grain sorghum, they have not yet been registered for this use [1116]. Recent work with glyphosate shows great promise and efforts are now being made for its registration [1117].

Sunflower

Sunflower is a fast-growing market for chemical harvest aids. Most of the land that is devoted to sunflowers in the summer months is generally planted to wheat and oats in the winter. Thus, the sooner the sunflowers can be harvested, the sooner these second crops can be planted. Sunflowers dry very slowly and mature long before they are dry enough to combine; these characteristics subject them to considerable mildew damage if bad weather occurs before harvest. Thus, use of paraquat as a harvest aid in sunflowers has been well received [1101, 1118].

Other Crops

Desiccants are used on several other legume crops other than soybeans. Temporary permission has been given for the use of dinoseb on lentils in the states of Idaho and Washington [1119]. Reports from Russia show that diquat is effective as a desiccant in trefoil [1120]; the treatment decreases the moisture content of the seed and increases the seed yield. Permission has also been given in the United States to apply paraquat to dry edible beans as a preharvest desiccant with the restriction that the foliage remaining after harvest not be fed to livestock.

Paraquat is registered in the United States for use as a harvest aid on sugarcane and guar [1101].

Table 20–1. Chemicals Used for Crop Desiccation

Chemical	Common Name (Trade Name)	Company	Crop	Reference
1,1′-dimethyl-4,4′-bipyridinium dichloride	paraquat (Gramoxone)	ICI	cotton	1094, 1096, 1108
			soybean	1098–1101, 1118
			rice	1108–1110
			guar	1101, 1108
			potato	1101, 1108
			sugarcane	1101, 1108
			sunflower	1101, 1108, 1118
			grain sorghum	1117
6,7-Dihydrodipyrido (1,2-a:2′,1′-c) pyrazidinium dibromide	diquat (Reglone, Reglox)	ICI	rice	1109, 1110
			potato	1111
			grain sorghum	1117
			trefoil	1120
2-Chloroethylphosphonic acid	ethephon (Ethrel)	Union Carbide	potato	1111
N-phosphonomethylglycine	glyphosate (Round-Up)	Monsanto	soybean	1098–1100
			rice	1109
			grain sorghum	1117
S,S,S-tributylphosphorothioate	– (DEF, Re-Green)	Mobay	cotton	1096
			rice	1102, 1108
2,5-Dichloro-4-nitrophenol	–	–	potato	1115
Ortho-arsenic acid	arsenic acid (Desiccant L-10, Zotox, Hi-Yield Desiccant H-10)	Commercial Chemical Pennwalt	cotton	1096
Sodium chlorate	sodium chlorate (Atlacide, Tumbleleaf)	ICI Kerr-McGee	cotton	1096
			rice	1103–1105, 1108, 1109
			trefoil	1120

Table 20–1. (continued)

Chemical	Common Name (Trade Name)	Company	Crop	Reference
Magnesium chlorate	magnesium chlorate (De-Fol-Ate, E-Z-Off)	–	rice	1102, 1108
2-(Ethylamino)-4-isopropylamino-6-methylthio-s-triazine	ametryn (Evik, Ametrex)	Ciba-Geigy	soybean	1100
2-Sec-butyl-4,6-dinitrophenol	dinoseb (Premerge 3, Naptro)	Dow Crystal Chemical Hoechst	rice lentils	1103, 1108 1119
7-Oxabicyclo(2,2,1)heptane-2,3-dicarboxylic acid	endothall (Accelerate, Aquathol, Des-i-cate, Hydrothol)	Pennwalt	rice	1102, 1104, 1108
Sodium chloroacetate	sodium chloroacetate	–	rice	1102–1105, 1107
Sodium pentachlorophenate	sodium pentachloro-phenate (Dowicide G-ST, Weedbeads)	Dow	rice	1103
Trichloroacetic acid	TCA (Dow Sodium TCA, Varitox)	Dow May & Baker	rice	1103

Chapter 21. Protection Against Herbicide Damage

A new concept in weed control involves the use of antidotes to protect crops from herbicidal injury. This concept stems from an observation made by Hoffman on a hot summer day: in a greenhouse filled with tomato plants that had been sprayed with 2,4-D analogs, all of the plants were dying from 2,4-D fumes except those that were treated with 2,4,6-trichlorophenoxyacetic acid [1121]. These latter plants appeared relatively normal. Since that time, Hoffman and many other investigators have found an array of antidotes for various herbicides [1122]. 2,4-D itself has an antidotal action for oats treated with barban [1123]. It was found that 2,4,6-T and MCPA also antidoted barban [1124], but these materials are too toxic in their own right to be used for the purposes desired.

Continued investigations by Hoffman led to the discovery of the first commercial herbicidal antidote, which was 1,8-naphthalic anhydride as an antidote for EPTC [1125]. Subsequently it has been found that 1,8-naphthalic anhydride is a most versatile seed treatment antidote [1124–1127]. It has protected corn against all of the thiocarbamates, dithiocarbamates, and chloroacetanilide herbicides on which it has been tried. It is useful also to antidote alochlor on rice and grain sorghum, to antidote molinate on rice, and to antidote barban on oats. N,N-diallyl-2,2-dichloroacetamide has been found to be even better in protecting corn from injury by EPTC and other thiocarbamate herbicides [1126, 1128–1132].

For some time there was little success in finding antidotes for herbicides that inhibit photosynthesis, especially those like atrazine that inhibit the Hill reaction. Separate studies have reported that certain diazosulfonates (especially dexon) [1133] and benzylhydrazones [1134] can act as antidotes to residual atrazine toxicity in soybeans.

A new safener for protecting sorghum from metolachlor injury has recently been announced. This compound is alpha(cyanomethoxamino)benzacetonitrile; its use permits effective control of a number of annual grasses in sorghum [1135, 1136].

Triallate, a herbicide widely used for wild oat control in wheat, must be superficially incorporated in the soil to prevent crop injury. Deeper incorporation gives more consistent control but also increases wheat injury. Effective antidotes should allow deeper incorporation of triallate, more consistent wild oat control, and no wheat injury. Two such antidotes have been reported: N,N-diallyl-2,2-dichloroacetanilide (R25788), and 5,6-dihydro-2-methyl-1,4-oxathin-3-carboxanilide (carboxin) [1137].

The herbicidal antidote concept was introduced by Hoffman in 1962. Within ten years, this new approach to enhance herbicide selectivity had already become an accepted agricultural practice. The continued interest in, and importance placed on, the safening of herbicides is shown by the number of patents issued in the United States during 1980 alone for compounds of this nature [1138–1146]. The crops involved include wheat [1144], corn [1137, 1144], grain sorghum [1140–1145], rice [1138–1142], soybean [1121, 1123], and cotton [1146]. The protection is against an array of herbicides including alkylthiocarbamates, thiocarbamates, chloroacetanilides, and triazines.

Fortunately, present herbicide safeners appear to be of very low toxicity to mammals [1147].

Table 21–1. Herbicide Antidotes

Herbicide	Antidote	Crop	Reference
4-amino-(1,1-dimethylethyl)-3-(methylthio)-1,2,4-triazin-5(4H)-one (metribuzin)	benzil hydrazone	soybeans	1134
2-tert-butylamino-4-ethylamino-6-methylthio-s-triazine (terbutryn)	benzil hydrazone	soybeans	1134
4-chloro-2-butynyl-*m*-chlorocarbanilate (barban)	1,8-naphthalic anhydride	oats	1124, 1126
2-chloro-2′,6′-diethyl-N-(butoxymethyl) acetanilide (butachlor)	2′-methoxy-3-phenacylidene phthalide	rice	1138
2-chloro-2′,6′-diethyl-N-(methoxymethyl) acetanilide (alachlor)	1,8-naphthalic anhydride	sorghum	1124
		rice	1124
	N,N-diallyl-2,2-dichloroacetamide	corn	1130
2-chloro-4-(ethylamino)-6-(1-cyano-1-methylethylamino)-1,3,5-triazine (cyanazine)	1-(phenylmethyl)-1H-pyrrole-2,3-dicarboxylic acid	grain sorghum	1143, 1145
2-chloro-4-(ethylamino)-6-(isopropylamino)-s-triazine (atrazine)	sodium p-(dimethylamino)benzene diazosulfonate (dexon)	soybeans	1133
	sodium p-methylbenzenediazosulfonate	soybeans	1133
	benzil hydrazone	soybeans	1134
2-chloro-N-(2-ethyl-6-methylphenyl)-N-(2-methoxy-1-methylethyl)-acetamide (metolachlor)	α-(cyanomethoximino)-benzacetonitrile	grain sorghum	1135, 1136
	N,N-diallyl-2,2-dichloroacetamide	corn	1130
	phenylglyoxylonitrile-2-oximecyanomethylether	millet	1140
5-(2,3-dichloroallyl)-diisopropylthiocarbamate (diallate)	methyl-N-ethyldithiocarbanilate	wheat	1144
		corn	1144
N-(3,4-dichlorophenylcarbamoyl)-N-methylglycine monohydrate	octamethylenediamine	cereals	1148
1,1′-dimethyl-4,4′-bipyridiniumdichloride (paraquat)	ferrous sulfate	wheat	1124
		oats	1124

Table 21–1. (continued)

Herbicide	Antidote	Crop	Reference
S-ethyldipropylthiocarbamate (EPTC)	1,8-naphthalic anhydride	corn	1124–1127
	1-dibromoacetyl-2,5-dimethylpyrrolidine	corn	1137
	N,N-diallyl-2,2-dichloroacetamide	corn	1126, 1129, 1131, 1132, 1142
S-ethylhexahydro-1H-azepine-1-carbothioate (molinate)	N-(2′)-benzimidazolyl-1,8-naphthalimid	rice	1139
	1,8-naphthalic anhydride	rice	1124
N-(2′-n-propoxyethyl)-2,6-diethylchloroacetanilide	phenylglyoxylonitrile-2-oximecyanomethylether	rice	1140–1142
		sorghum	1140–1142
5-(2,3,3-trichloroallyl)-N,N-diisothiocarbamate (triallate)	N,N-diallyl-2,2-dichloroacetamide	wheat	1137
	5,6-dihydro-2-methyl-1,4-oxathin-3-carboxanilide	wheat	1137
	methyl-N-ethyldithiocarbanilate	wheat	1144
		corn	1144

Chapter 22. Increase of Herbicide Absorption and Translocation

The first work of this nature was by Binning and his colleagues [1149] who showed that pre-treatment with ethephon increased the basipetal translocation of the herbicide dicamba in wild garlic.

Canada thistle is a difficult weed to control even with the most effective herbicides. This is due to the lack of herbicidal movement to the roots in quantities large enough to kill the root buds. Ethephon was found to increase the effectiveness of both dicamba and 2,4-D (the herbicides of choice for the control of Canada thistle) when applied in combination with either of these herbicides [1150]. More recent work has shown a similar effect for chlorflurenol when applied with dicamba on Canada thistle [1151].

Chapter 23. Toxicology, Environmental and Human Safety

For regulatory purposes, plant growth regulators are considered in the same category as pesticides and the same rigid requirements apply. Because plant growth regulators sometimes are applied closer to harvest than are pesticides, the actual practical requirements for their safe use become even more stringent.

Anderson and his co-workers conducted mutagenicity tests using four different microbiological systems to evaluate the ability of more than one hundred pesticides to cause point mutations in the microorganisms used in these tests [1152]. The tests employed were a modified Ames test and the use of bacteriophage mutants. Included in the tests were the following pesticides that are sometimes used as plant growth regulators: cacodylic acid, 4-CPA, 4-CPP, 2,4-D, dinoseb, dimequat, diuron, DNOC, endothal, MH, monuron, paraquat, and 2,3,6-trichlorobenzoic acid. None of these materials appeared to cause point mutation of the microbial systems used. Known mutagens included as standards were 5-bromouracil and 2-aminopurine. Several commonly occurring materials including sucrose, fertilizer, aspirin, and pepper were found to have the same mutagenic frequencies as those of the pesticides under evaluation; the authors considered this to be evidence that these pesticides lacked mutagenic properties. Although Northrop [1153] had earlier implicated maleic hydrazide as a mutagenic chemical in the microbiological systems he employed, the studies of Anderson and his co-workers [1152] do not confirm Northrop's conclusions. Northrop's system involved the induction of virus production.

The question of the safety of maleic hydrazide on crops was raised in the United States during the early 1970's and a Rebuttable Presumption Against Registration (RPAR) was issued by the Environmental Protection Agency in October of 1977 [1154]. A review of maleic hydrazide and the question of its safety was given by an official of the U.S. Food and Drug Administration [1155]. He listed among its hazardous possibilities

(a) that it is a carcinogen [1156, 1157];
(b) that it causes chromosome breakage in mice [1158];
(c) that it causes chromosomal aberrations in plants [1159–1161];
(d) that plants may break it down to hydrazine, a known carcinogen [1162];
(e) that residues of maleic hydrazide in potatoes decreased fertility in rats [1163]; and
(f) that the chemical retards growth and is toxic to amphibians [1164].

More recent studies [1165] found that maleic hydrazide in amounts far in excess of that encountered from residues in food had no apparent effect on liver microsomal enzymes in rats; the evidence from this investigation also suggests that maleic hydrazide would not be expected to alter the metabolism of other chemicals that might be present in biological systems. Still more recent work in Japan [1166] done with Chinese hamster cells *in vitro* showed maleic hydrazide in itself to have weak inducibility of cytotoxicity but positive cytogenetic effects on the cells. The conclusions are that maleic hydrazide is a hazardous chemical in the environment and warrants additional studies, a conclusion agreed with by the U.S. Food and Drug Administration.

2,4-D and its related phenoxy acids used as herbicides have moderately acute toxicity and are moderate in their local effects upon the skin or eyes [1167]. Both chronic and subchronic toxicity studies indicate that none of these phenoxy acids exhibits a significant potential chronic toxicity. Several studies support the conclusion that 2,4-D is not carcinogenic [1167–1170]. Recent cytogenetic studies in Sweden [1171] conclude that in practice the lack of penetration of 2,4-D into cells indicates that it does not constitute a cytogenetic hazard to man. Contemporary work from Finland [1172, 1173] suggests that the central nervous system is involved in the toxicity of chlorophenoxyacetic acid administered in high dosages.

Because of the widespread use of ethephon and because some organohalides have been linked to cancer in both human and animal studies [1174], both the carcinogenicity and the co-carcinogenicity of ethephon were assessed by Strain A mouse pulmonary adenoma bioassay [1175]. The results show that not only does ethephon not pose a carcinogenic threat to man, but it may actually confer some protection against environmental carcinogens. In these studies, ethephon suppressed the spontaneous development of lung tumors in Strain A mice and inhibited the development of lung tumors formed in response to urethane, an active carcinogen in this bioassay.

Mammalian studies with ingested chlormequat showed it to be principally eliminated unchanged in the urine with lesser amounts eliminated in the feces and in respiratory gases [1176]. Residual amounts in various tissues were in very low concentration.

Of the many defoliants used on the cotton crop in the United States, sodium chlorate is the only harvest aid chemical which is not at least a proposed candidate for RPAR by the U.S. Environmental Protection Agency [238]. Arsenic acid became an official candidate for RPAR in 1978 [237]. There had been questions as to the safety of the organophosphorous compounds S,S,S-tributylphosphorotrithioate and tributylphosphorotrithioite; these questions revolve around delayed neurotoxic effects, late acute effects, and cholinergic effects of these two materials [1177–1179].

Although paraquat is a relatively safe compound when used as directed, significant toxicity may result from inappropriate use [1180].

Although paraquat is a lung irritant [1181] and its pulmonary toxicity is well documented [1182], inhalation of the spray has apparently never been associated with human fatality [1183]. Poisoning with concentrated solutions of paraquat is a highly lethal event [1183]. The first death resulting from paraquat ingestion was reported in 1966 [1184]. Worldwide deaths from paraquat numbered 124 in 1971, about equally divided between suicides and accidental ingestion, usually the result of drinking from unlabelled containers [1181]. By 1975, over 500 fatalities had been reported [1180]. Recently, a great deal of attention has been focused on the toxicity of paraquat due to its contamination of marijuana through an effort by the Mexican government to eradicate the plant [1185].

Therapy for human paraquat consumption consists of removal of the poison by gastric lavage, administration of Fuller's earth or Bentonite suspensions to absorb the chemical, and osmotic diuresis with mannitol to aid excretion [1181]. Paraquat does not constitute a very dangerous inhalation hazard under customary conditions of use, but protective equipment, particularly respiratory, should be used when contamination might occur [1181]. Drifting spray should be avoided. Paraquat has not been shown to be mutagenic, carcinogenic, or teratogenic in animals [1181].

In spite of the large number of deaths due to ingestion of all types of bipyridylium quaternary ammonium compounds, these materials have a good safety record in agriculture [1186].

Recent work in Dow Chemical's toxicology research laboratories showed indoleacetic acid, the naturally-occurring plant hormone, to be teratogenic in mice and rats at 500 mg/kg/day, inducing cleft plate in both species at this dose level [1187]. The compound gave no evidence of maternal toxicity at this level and was not teratogenic below the 500 mg/kg/day dose. Because it is one of the most generally used herbicides in Europe, several toxicological studies have been conducted with 2-methyl-4-chlorophenoxyacetic acid (MCPA). In none was it considered to be highly toxic [1188–1190].

In a 90-day feeding study with naphthaleneacetic acid in rats, a no-effect level of 1000 parts per million would have been indicated by the study on the basis of conventional criteria [1191]. However, in view of liver glycogen depletion at all levels tested, a no-effect level could not be established. In December of 1980, the U.S. Environmental Protection Agency expanded the tolerances for NAA and its ethyl ester based on a number of toxicology studies, a three-generation mouse study, a two-year rat feeding study, a rat teratology study, a rabbit eye irritation study, and several mutagenicity tests. An *Acceptable Daily Intake* (ADI) of 0.005 mg/kg/day was established based on a 90-day feeding study with a 2,000-fold safety factor [1192].

Chapter 24. Summary

Normal plant growth and development is controlled by chemicals produced by the plant itself (endogenous) plant hormones. It is generally accepted that synthetic plant-growth regulators produce their effects through changing the endogenous level(s) of the naturally-occurring hormones, thus allowing a modification of growth and development in the desired direction and, it is hoped, to the desired degree.

The discovery of 2,4-D as a herbicide during World War II precipitated the greatest single advance in the science of weed control and one of the most significant in agriculture [1193]. The history of 2,4,5-T has closely paralleled that of 2,4-D since 1941 when Pokorny [1194] reported the synthesis of both compounds. In 1942, Zimmerman and Hitchcock [1195] reported that at extremely low dosages, 2,4-D and its derivatives were active in inducing formative effects in plants; they were the first to establish certain substituted phenoxy compounds as plant growth substances. In 1944, Hammer and Tukey [1196] reported the effectiveness of 2,4-D and 2,4,5-T as herbicides on bindweed. In the same year, reports told of the selectivity of 2,4-D for removal of broadleaf weeds in lawns [1197] and in the field [1198] without injury to established grasses.

An ever-increasing role for growth regulators is assured by the certain increase in the cost of energy, by the continual decrease in prime productive land as it is converted to urban and industrial uses, and by the need to double the world's food supply by the end of the 20th century. The need to increase agricultural production at an unprecedented rate almost dictates that plant growth regulators will become a major contributor in reaching the desired goals.

Manipulating crop production processes with chemicals may be one of the most important advances to be achieved in agriculture. The levelling-off of the gains achieved from the use of fertilizers and pesticides, the need for increased world food production, and the need for more efficient use of available energy – all three point to the usefulness of controlling plant growth processes to aid in the efficiency and the output of agriculture.

The diversity of the effects of plant growth regulators is a complicating factor and one that has practical economic overtones; nevertheless, this diversity is also an outstanding positive feature – there seems to be virtually no limit to the number of ways in which growth regulators may be used. These ways range from the more obvious visual effects such as change in shape, size, or growth rate, to those more subtle changes that affect metabolism and thus result in a change in both quality and quantity of the desired economic product.

The specific uses of growth regulators for horticultural purposes are usually somewhat different from those for agronomic crops. The approach to economics as well as the basis for economic justification is also different, although these differences are primarily one of degree. For horticultural purposes, the control of growth and development is dominant, whereas the control of metabolism is the major factor for agronomic crops.

Although less emphasis has been placed to date on the potential for overcoming environmental stresses, success in this area has great practical implications. Among the studies suggested for concentrated effort in the near future would be use of growth regulators: (a) to allow a crop to be moved farther north or south according to its ability to withstand more cold or less water, or (b) to change its time pattern so that the crop can mature and be harvested before adverse conditions are imposed. Success already achieved promises substantial rewards from such efforts.

Concentration of research on specific crops should give added benefits not always attainable by industry, commonly forced by economics to take a broader look at its target areas. The success with crops where such major efforts have been aimed is dramatically demonstrated for rubber, sugarcane, apples, and cotton.

Historically, man has been able to improve his crops (a) by alterations in their genetic composition, (b) by changing the environment in which the crop grows, and (c) by certain manipulative techniques such as pruning and grafting. There now is no question that a fourth method for crop improvement can be added – that of using plant growth regulators. In spite of limitations, the use of plant growth regulators brings new opportunities to agriculture and to horticulture in removing or circumventing many of the barriers imposed by genetics and the environment. Shortening the time span necessary to accomplish certain objectives set by breeding programs through the use of plant growth regulators has, in some instances, been most gratifying. If the limitations, constraints, and potential side effects of the use of plant growth regulators are kept in mind and in perspective, and if the proper emphasis and support given to this field of investigation, society can expect great progress.

Further, if it is kept in mind that both breeding programs and research on plant growth regulators have the same primary objective – the modification of crop performance for the benefit of society – and if they are used in a combination or in a supplementary manner rather than as alternatives, this objective will be achieved sooner.

How high can yields go? The factor that sets the upper limit on potential yield is the quantity of energy that crop tissues capture from the sun. Energy relationships play an important role in establishing the theoretical, as well as the practical, upper limits of crop productivity. The proportion of captured solar energy that goes into oil, protein, and carbohydrate also affects the final yield potential. Partitioning or shifting the desired product into the "economic" portion of the crop is one of the areas of great potential for plant growth regulators, as has been dramatically shown in the case of sugarcane ripeners. Agriculture is one of the oldest industries that has attempted to maximize the use of solar energy. Although considerable progress has been made, there is much room for improvement [1199]. Among areas of improvement are some in which the use of plant growth regulators should fit.

In this book the many plant processes that can be modified by the use of plant growth regulators have been discussed, as have the specific compounds that bring about these modifications. As the need for using plant growth regulators is increasingly appreciated and emphasis on them increases, we can expect (a) new types of uses for plant growth regulators, (b) replacement of many of the regulators now being used by others that are more effective, and (c) an increase in the economic return, both immediate and longrange, from their use.

Many investigators involved in agricultural research have predicted that sales of plant growth regulators will be equal to, or outstrip, herbicides in the "reasonable" future [1200]. One of the more recent comments on this point was made by Hardy [1201], "By 2000 A.D., I predict that plant growth regulators, primarily as yield enhancers, secondarily as quality improvers and production process facilitators, will become as important as fertilizers or plant protectant chemicals for world crop production".

As progress continues to be made in the basic understanding of the physiology and the genetics of plant growth and development, there will be an increased efficiency in finding plant growth regulators to perform in the desired function, discoveries based more and more on designing the type of molecule to achieve the desired effect rather than our past dependence on the empirical approaches.

Table 24–1. Cited Chemicals with Plant Growth Regulator Activity

Chemical Name	Common Name	Designation or Code	Trade Name	PGR Activity	Other Activity	Chapter Location
acetylene	acetylene	–	–	flowering inducer	–	1, 4
alkenylsuccinic acid	–	–	–	anti-transpirant	–	14
p-aminobenzenesulfonyl urea	–	–	–	sugarcane ripener	–	17
aminoethoxyvinylglycine	AVG	–	–	fruit set enhancer fruit ripening inhibitor sex changer	–	7, 17, 18
2-amino-4-methoxy-3-butenoic acid	–	–	–	growth retardant	–	8
2-amino-6-methylbenzoic acid	–	–	–	fruit set enhancer growth retardant sugarcane ripener fruit ripener	–	7, 8, 17
aminomethylphosphonic acid	–	AMPA	–	sugarcane ripener	–	17
6-aminopenicillanic acid	–	6-APA	–	sugarcane ripener	antibiotic nucleus	17
3-amino-1,2,4-triazole	amitrole	–	Amino Triazole	insect anti-feedant	herbicide	13
4-amino-3,5,6-trichloropicolinic acid	picloram	–	Tordon	sugarcane ripener peanut stimulator	herbicide	17
ammonium isobutyrate	–	AIB	–	sugarcane ripener	fungicide bactericide	17
o-arsenic acid	arsenic acid	–	Desiccant L-10 Zotox	desiccant	–	20
bacitracin	bacitracin	–	–	sugarcane ripener	antibiotic	17
N-benzoyl-N-(3,4-dichlorophenyl)-aminopropionic acid	–	–	–	sugarcane ripener	–	17

Table 24–1. (continued)

Chemical Name	Common Name	Designation or Code	Trade Name	PGR Activity	Other Activity	Chapter Location
6-benzylaminopurine	6-benzyl-adenine	BAP	–	fruit shape changer tillering agent sex changer senescence delayer	–	7, 12, 17, 18, 19
6-benzylamino-9-(tetrahydropyran-2-yl)-9H-purine	–	PBA	–	tillering agent	–	12
5-bromo-3-sec-butyl-6-methyl uracil	bromacil	–	Hyvar X	sugarcane ripener	herbicide	17
2-sec-butyl-4,6-dinitrophenol	dinoseb	DNBP	Dinitro Premerge 3 Amaize Spark Basanite	corn yield enhancer desiccant	herbicide	17, 20
chlorinated camphene	toxaphene	–	Camphoclor Cristoxo Phenacide Motox	cotton boll promoter	insecticide	4
2-chlorobenzoic acid	–	–	–	sugarcane ripener	–	17
1-(2-chlorobenzyl)-3-carboxy-4,6-dimethyl-pyrid-2-one	–	–	–	gametocide	–	5
2-chloro-2′,6′-diethyl-N-(methoxymethyl)-acetanilide	alachlor	–	–	anti-transpirant	herbicide	14
2-chloroethanethiophosphonic acid dichloride	–	–	–	fruit ripener	–	17
2-chloroethylaminodi(methylphosphonic) acid	–	–	–	sugarcane ripener	–	17
β-chloroethylmethyl-bis-benzyloxysilane	–	CGA-15281	–	fruit thinner	–	6
2-chloro-N-(2-ethyl-6-methylphenyl)-N-(2-methoxy-1-methylethyl)acetamide	–	CGA-24705	–	growth retardant	–	8

Table 24–1. (continued)

Chemical Name	Common Name	Designation or Code	Trade Name	PGR Activity	Other Activity	Chapter Location
2-chloroethylphosphonic acid	ethephon	68-250 CEPA	Ethrel Cepha Florel	tillering agent sugarcane ripener flowering inducer color enhancer latex stimulator gametocide abscission agent fruit thinner growth retardant herbicide mover pigment former fruit ripener sex changer desiccant	–	4, 5, 6, 8, 12, 16, 17, 18, 20, 22
(2-chloroethyl)trimethylammonium chloride	chlormequat	CCC	Cycocel	sugarcane ripener growth retardant lodging reducer germination inhibitor flower bud stimulator pigment former sex changer senescence delayer fruit set enhancer insect controller anti-transpirant	–	3, 4, 7, 8, 13, 14, 15, 16, 17, 18, 19
2-chloroethyl-tris (2′-methoxyethoxy)silane	–	CGA-13586	Alsol	abscission agent	–	6
3-chloroisopropyl-N-phenyl carbamate	chlorpropham	Chloro-IPC CIPC	Sprout-Nip Spud-Nic Beet-Kleen	fruit thinner	herbicide	6

Table 24–1. (continued)

Chemical Name	Common Name	Designation or Code	Trade Name	PGR Activity	Other Activity	Chapter Location
2-chloromercuri-4,6-dinitrophenol	–	–	–	anti-transpirant	–	14
5-chloro-3-methyl-4-nitro-1H-pyrazole	–	–	Release	abscission agent	–	6
4-chloro-2-methylphenoxyacetic acid	MCPA metaxon	MCPA	Agroxone Rhonox Weedar	herbicide antidote	herbicide	21
4-chlorophenoxyacetic acid	4-CPA	–	–	fruit set enhancer	–	7
α-(p-chlorophenoxy)isobutyric acid	–	–	–	fruit ripening inhibitor	–	17
3-chlorophenoxy-α-propionamide	–	–	–	fruit thinner	–	6
1-(p-chlorophenyl)-1,2-dihydro-4,6-dimethyl-2-oxonicotinic acid, sodium salt	–	RH-531	–	gametocide soybean stimulator	–	5, 17
3-(p-chlorophenyl)-1,1-dimethyl urea	monuron	–	Telvar Monurex	flowering inhibitor	herbicide	4
3-(p-chlorophenyl)-6-methoxy-s-triazine-2,4-(1H,3H)-dione, triethanolamine salt	–	DPX-3778	–	gametocide	–	5
2-(4-chlorophenylthio)triethylamine hydrochloride	CPTA	–	–	pigment former	–	16
2-chloro-4-quinoline carboxylic acid	–	–	–	gametocide	–	5
5-chloro-2-thenyl-tri-n-butylphosphonium chloride	–	CHE-8728	–	sugarcane ripener	–	17
N′-(4-chloro-o-tolyl)-N,N-dimethyl formamidine	chlordimeform	C-8514	Acaron Bermat Fundal	cotton boll retainer	insecticide acaricide	4
α-(cyanomethoxamino)benzacetonitrile	–	–	–	herbicide antidote	–	21
3-cyclohexene-1-carboxylic acid	tetrahydro-benzoic acid	–	–	sugarcane ripener	–	17

Table 24–1. (continued)

Chemical Name	Common Name	Designation or Code	Trade Name	PGR Activity	Other Activity	Chapter Location
α-cyclopropyl-α-(p-methoxyphenyl)-5-pyrimidine methanol	ancymidol	EL-531	A-Rest	growth retardant	–	8
N,N-diallyl-2,2-dichloroacetamide	–	–	–	herbicide antidote	–	21
N,N-diallyl-2,2-dichloroacetanilide	–	R25788	–	herbicide antidote	–	21
2′,4′-dichloro-1-cyanoethanesulfonanilide	–	–	–	chlorophyll remover	–	16
dichlorodiphenyltrichloroethane	DDT	–	Anofex Arkotine Dedelo	cotton boll promoter	insecticide	4
2,4-dichloro-5-fluorophenoxyacetic acid	–	–	–	gametocide	–	5
2,3-dichloroisothiazole-5-carboxylic acid	–	–	–	insect controller	–	13
2,3-dichloro-6-methyl-benzoic acid	–	–	–	sugarcane ripener	–	17
2,3-dichloro-2-methyl propionic acid, sodium salt	–	FW 450	Mendok	gametocide	–	5
2,5-dichloro-4-nitrophenol	–	–	–	desiccant	–	20
2,4-dichlorophenoxy acetic acid	2,4-D	–	Plantgard Agrotect Weed-B-Gon	fruit drop controller latex stimulator disease controller insect controller senescence delayer root inducer flowering inducer fruit set enhancer fruit ripener herbicide antidote	herbicide	2, 4, 7, 13, 15, 17, 19, 21
2-(3,4-dichlorophenoxy) triethylamine	–	–	–	latex stimulator	–	12

Table 24–1. (continued)

Chemical Name	Common Name	Designation or Code	Trade Name	PGR Activity	Other Activity	Chapter Location
3-(3,4-dichlorophenyl)-1,1-dimethyl urea	diuron	–	Karmex Krovar Urox Dynex	flowering inhibitor	herbicide	4
N,N-diethylnonylamine	–	–	–	pigment former	–	16
2,3-dihydro-5,6-dimethyl-1,4-dithiin-1,1,4,4-tetraoxide	–	–	–	desiccant	–	20
6,7-dihydropyrido(1,2-a:2′,1′-c) pyrazidinium dibromide	diquat	–	Reglone Reglox Dextrone	flowering inhibitor desiccant	herbicide	4, 20
5,6-dihydro-2-methyl-1,4-oxathin-3-carboxanilide	carboxin	–	–	herbicide antidote	–	21
2,3-dihydro-5-(4-methylphenyl)-6-phenyl-1,4-oxathiin-4-oxide	–	–	–	growth retardant	–	8
1,2-dihydro-3,6-pyridazinedione	maleic hydrazide	MH	MH-30 Sprout Stop Sucker-Stuff De-Cut	growth retardant flowering inhibitor gametocide axillary bud controller insect controller anti-ripener	–	3, 4, 5, 8, 9, 13, 17
N-(2,3-dihydroxy-1-propyl)-N-phosphonomethylglycine, disodium salt	–	–	–	sugarcane ripener	–	17
diisobutylphenoxyethoxyethyldimethylbenzylammonium chloride	–	–	Hyamine 1622	sugarcane ripener	biocide	17
O,S-dimethylacetylphosphoramidothioate	acephate	Ortho-12420	Orthene Ortran	cotton flowering stimulator	insecticide	4

Table 24–1. (continued)

Chemical Name	Common Name	Designation or Code	Trade Name	PGR Activity	Other Activity	Chapter Location
2-(β-dimethylamino-ethoxy)-4-(3′,4′-dichloro-phenyl)-thiazole hydrochloride	–	–	–	sugarcane ripener	–	17
dimethylamino maleamic acid	–	–	–	flower bud stimulator	–	4
[4-(dimethylamino)phenyl]-diazenesulfonic acid, sodium salt	fenaminosulf	Bay 5072 Bay 22555	Dexon	herbicide antidote	fungicide	21
dimethylarsinic acid	cacodylic acid	–	Phytar 560	defoliant sugarcane ripener desiccant	herbicide	6, 17, 20
1,1′-dimethyl-4,4′-bipyridinium dichloride	paraquat	–	Gramoxone	oleoresin stimulator desiccant	herbicide	1, 7, 20
N,N-dimethylglycine	–	–	–	sugarcane ripener	–	17
4,4-dimethylmorpholinium chloride	–	–	–	cotton boll suppressor insect controller	–	4, 13
O,O-dimethyl-S-[(4-oxo-1,2,3-benzotriazin-3(4H)-yl)-methyl]phosphorodithioate	azinphosmethyl	Bay 17147	Guthion Garfene	cotton flower and boll promoter	insecticide	4
3-(2-[3,5-dimethyl-2-oxocyclohexyl]-2-hydroxyethyl)-glutarimide	cycloheximide	–	Acti-Aid	abscission agent sugarcane ripener fruit ripening inhibitor	antifungal antibiotic	6, 17
2-(1-[2,5-dimethylphenyl]-ethylsulfonyl)-pyridine-N-oxide	–	–	–	growth retardant	–	8
1,1-dimethyl-piperidinium chloride	mepiquat	BAS 0830	Pix	growth retardant disease controller	–	8, 13
N-2,4-dimethyl-5-(trifluoromethyl)-sulfonyl-aminophenylacetamide	mefluidide	MBR 12325	Embark	sugarcane ripener growth retardant	herbicide	8, 17

Table 24–1. (continued)

Chemical Name	Common Name	Designation or Code	Trade Name	PGR Activity	Other Activity	Chapter Location
4,6-dinitro-2-sec-butyl phenol	dinoseb	DNBP	Dinitro Premerge 3 Basanite Amaize Spark	corn yield enhancer desiccant	herbicide	17, 20
dinitrocyclohexophenol	–	–	–	fruit thinner	–	6
4,6-dinitro-orthocresylate, sodium salt	DNOC	–	Elgetol Chemsect Sinox	fruit thinner	insecticide fungicide herbicide	6
2,4-dinitrophenol	DNP	–	Chemox	anti-transpirant	insecticide fungicide	14
ethanedialdioxime	glyoxime	–	Pik-Off	abscission agent	–	6
ethanolamine-p-nitrobenzene-sulfonylurea	–	MON-814	–	growth retardant	–	8
ethene	ethylene	–	–	flowering inducer latex stimulator fruit ripener sex modifier defoliant abscission agent de-greener		1, 4, 6, 16, 17
4-ethoxy-1(p-tolyl)-s-triazine-(2,6)(1H:3H) dione	–	–	–	growth stimulator	–	8
2-(ethylamino)-4-isopropylamino-6-methyl-thio-s-triazine	ametryn	G-34162	Evik	desiccant	herbicide	20
ethyl-5-chloro-1H-indazol-3-acetic acid ethyl ester	–	–	–	fruit thinner	–	6

Table 24–1. (continued)

Chemical Name	Common Name	Designation or Code	Trade Name	PGR Activity	Other Activity	Chapter Location
ethyl-5(4-chlorophenol)-2H-tetrazole-2-yl-acetate	–	–	–	chemical pruner tree shaper	–	10, 11
N-(m-fluorobenzyl)phthalimide	–	–	–	gametocide	–	5
4-fluoro-2,6-dichlorophenoxyacetic acid	–	–	–	gametocide	–	5
6-furfurylaminopurine	kinetin	–	–	dormancy breaker germination stimulant disease controller	–	3, 13
hexadecyltrimethylammonium bromide	cetyltrimethyl-ammonium bromide	CTAB	Cetrimide	sugarcane ripener	biocide	17
hydroxydimethylarsine oxide	cacodylic acid	–	Phytar 560	defoliant desiccant sugarcane ripener	herbicide	6, 17, 20
1-hydroxy-1,1-ethane diphosphonic acid	–	–	–	sugarcane ripener	–	17
β-hydroxyethylhydrazine	BOH	–	–	flowering inducer	–	4
4-hydroxy-3-methoxybenzaldehyde	vanillin	–	–	sugarcane ripener	flavoring	17
8-hydroxyquinoline	–	–	–	anti-transpirant	–	14
1-hydroxytriacontane	triacontanol	–	–	growth stimulator	–	8
imidodicarbonic diamide	carbamyl urea	–	Biuret	sugarcane ripener	–	17
3-indoleacetic acid	IAA	–	–	cell enlarger disease controller anti-transpirant fruit ripening inhibitor	–	1, 7, 13, 14, 17
3-indolebutyric acid	IBA	–	Hormodin	root inducer fruit set enhancer	–	2, 7

Table 24–1. (continued)

Chemical Name	Common Name	Designation or Code	Trade Name	PGR Activity	Other Activity	Chapter Location
iodoacetic acid	–	–	–	abscission agent	–	6
isochlortetracycline	isoaureomycin	–	–	sugarcane ripener	antibiotic degradation product	17
laurylmercaptotetrahydropyrimidine	–	–	–	sugarcane ripener	fungicide	17
manganese chlorate	–	–	–	defoliant desiccant	–	6, 20
β-mercaptovaline	penicillamine	–	Cuprimine	sugarcane ripener	arthritis drug	17
2-(p-methoxybenzyl)-3,4-pyrrolidine-diol-3-acetate	anisomycin	–	Flagecidin	sugarcane ripener	antibiotic	17
N-[(4-methoxy-6-methylamino-1,3,5-triazin-2-yl)-aminocarbonyl]benzene sulfonamide	–	–	–	sugarcane ripener	–	17
1-methyl-3-carboxy-4,6-dimethylpyrid-2-one	–	–	–	gametocide	–	5
5-methyl-7-chloro-4-ethoxycarbonylmethyl-2,1,3-benzothiadiazole	–	TH-6241	–	sex changer	–	18
methyl-2-chloro-9-hydroxyfluorene-9-carboxylate	chlorflurecol-methyl	CF-125 IT-3299	Po-San Chlorflurecol	herbicide mover fruit ripener fruit set enhancer growth retardant tillering agent insect controller anti-transpirant	–	7, 8, 12, 13, 14, 17, 22
2-methyl-4-chlorophenoxyacetic acid	MCPA metaxon	MCPA	Agroxone Rhonox Weedar	herbicide antidote	herbicide	21

Table 24–1. (continued)

Chemical Name	Common Name	Designation or Code	Trade Name	PGR Activity	Other Activity	Chapter Location
methyldecanoate	–	–	–	chemical pruner	–	10
methyl-3,6-dichloro-o-anisate	disugran	60-CS-16	Racuza	sugarcane ripener	–	17
methyl-2,7-dichloro-9-hydroxyfluorene-9-carboxylate	dichlorflurenol methyl	IT-3353	–	growth retardant	–	8
methyl-N′,N′-dimethyl-N-[(methylcarbamoyl)oxy]-1-thiooxamimidate	oxamyl	–	Vydate	fruit thinner	insecticide nematicide	6
2,3:4,6-bis-(1-methylethylidene)-O-(L)-xylo-2-hexulofuranosonic acid, sodium salt	dikegulac	–	Atrinal	growth retardant chemical pruner	–	8, 10
methyl-9-hydroxyfluorene-9-carboxylate	flurenol	IT-3235	–	growth retardant	–	8
7-methyl indole	–	PP-757	–	sugarcane ripener	–	17
3-(2-methylphenoxy)pyridazine	–	H-722	Credazine	sugarcane ripener	–	17
2-methyl-1-propanol	isobutanol	–	–	sugarcane ripener	solvent	17
methylsulfanil-yl-carbamate	asulam	MB-9057	Asulox	sugarcane ripener	herbicide	17
naphthaleneacetamide	NAD NAAm	–	Amid-Thin W	fruit thinner root inducer disease controller	–	6, 13
1-naphthaleneacetic acid	NAA	–	Fruitone-N Tre-Hold Transplantone Niagara Stik	latex stimulator fruit thinner flower inducer root inducer preharvest fruit drop preventer fruit set enhancer disease controller sex changer senescence delayer	–	4, 6, 7, 13, 17, 18, 19

Table 24–1. (continued)

Chemical Name	Common Name	Designation or Code	Trade Name	PGR Activity	Other Activity	Chapter Location
1,8-naphthalic anhydride	–	–	–	herbicide antidote	–	21
2-naphthoxyacetic acid	BNOA NOA	–	–	fruit set enhancer disease controller	–	7, 13
1-naphthyl-N-methyl carbamate	carbaryl	–	Sevin	fruit thinner	insecticide	6
N-1-naphthylphthalamic acid	naptalam	ACP 322 NPA-3	Alanap	fruit thinner insect anti-feedant	herbicide	6, 13
7-oxabicyclo(2,2,1)heptane-2,3-dicarboxylic acid	endothal	–	Ripenthol	growth retardant sugarcane ripener desiccant	herbicide	8, 17, 20
pentachlorophenol, sodium salt	penta	PCP	Dowicide	desiccant	wood preservative	20
n-pentanoic acid	n-valeric acid	–	–	sugarcane ripener	–	17
6-phenoxyacetamido penicillanic acid	penicillin V	Pen-V	Pen-Vee	sugarcane ripener	antibiotic	17
phenoxyacetic acid	POA	–	–	fruit set enhancer	–	7
N-(2-phenoxyethyl)-N-propyl-1H-imidazole-1-carboxamide	–	BTS 34-273	–	sugarcane ripener	–	17
2-(1-phenylethylsulfonyl) pyridine-N-oxide	–	–	–	growth retardant	–	8
phenylmercuric acetate	PMA	–	–	anti-transpirant	–	14
N-phenylphosphinylmethyliminodiacetic acid-N-oxide	–	–	–	sugarcane ripener	–	17
N-phenylsulfonamido-N-phosphonomethyl-glycine	–	–	–	sugarcane ripener	–	17
3-phenyl-1,2,4-thiadiazol-5-yl-thioacetic acid	–	–	–	tree shaper	–	11
N-phenyl-N′-1,2,3-thiadiazol-5-yl-urea	–	–	–	defoliant	–	6

Table 24–1. (continued)

Chemical Name	Common Name	Designation or Code	Trade Name	PGR Activity	Other Activity	Chapter Location
phosphonic acid, (2,2,2-trichloro-1-hydroxy-ethyl)-bis-[2-(2-hydroxypropoxy)-1-methylethyl]ester	–	–	–	sugarcane ripener	–	17
phosphonomethylglycine	glyphosate	MON-139 MON-8000	Round-Up Polado	sugarcane ripener growth retardant fruit ripener desiccant	herbicide	8, 17, 20
N,N-bis-(phosphonomethyl) glycine	glyphosine	MON-845	Polaris	sugarcane ripener growth retardant	–	8, 17
phosphonomethylglycine, calcium salt	–	MON-464	–	growth retardant	–	8
N-phosphonomethyliminodiacetic acid	–	MON-820	–	growth retardant	–	8
polychlorocarboxylic acid esters	–	–	–	gametocides	–	8
poly[oxyethylene(dimethylimino)ethylene-(dimethylimino)ethylene dichloride]	–	–	Bualta	sugarcane ripener	–	17
potassium 3,4-dichloroisothiazole-5-carboxylate	–	–	–	growth stopper defoliant	–	6
indent						
propyl-3-t-butylphenoxy acetate	–	M&B 25-105	–	tree shaper	–	14
salicylaldoxime	–	–	–	anti-transpirant	–	14
sodium chlorate	–	–	–	defoliant desiccant	–	6, 20
sodium chloroacetate	–	–	–	desiccant	–	20

Table 24–1. (continued)

Chemical Name	Common Name	Designation or Code	Trade Name	PGR Activity	Other Activity	Chapter Location
succinic acid-2,2-dimethyl hydrazine	daminozide	SADH	Alar Kylar B-Nine B-9	germination inhibitor growth retardant flowering inducer flowering delayer fruit set enhancer fruit set reducer disease controller anti-transpirant pigment former fruit ripener senescence delayer	–	3, 4, 7, 8, 13, 14, 15, 16, 17, 19
tetrahydrofurfuryl isothiocyanate	–	–	–	soybean stimulator	–	17
tetrahydrofuroic acid hydrazide	–	–	–	sugarcane ripener	–	17
(22R,23R,24S)-2α-3α,22,23-tetrahydroxy-24-methyl-6,7-s-5α-cholestano-6,7-lactone	brassinolide	–	–	growth stimulator	–	8
1,1,5,5-tetramethyl-3-dimethyl amino-dithiobiuret	–	–	–	fruit thinner	–	6
n-meta-tolyl-phthalamic acid	–	–	–	fruit set enhancer	–	7
1-triacontanol	–	–	–	growth stimulator	–	8
1,2,4-triazine-3,5-(2H,4H)-dione	6-azauracil	–	–	sugarcane ripener	–	17
tributyl[(5-chloro-2-thienyl)methyl]phosphonium chloride	–	–	–	disease controller	–	13
tributyl (5-chloro-2-thienyl) phosphonium chloride	–	–	–	soybean stimulator	–	17
tributyl-2,4-(dichlorobenzyl) phosphonium chloride	chlorphonium	–	Phosphon	growth retardant flower bud stimulator insect controller	–	3, 4, 13

Table 24–1. (continued)

Chemical Name	Common Name	Designation or Code	Trade Name	PGR Activity	Other Activity	Chapter Location
S,S,S-tributylphosphorotrithioate	DEF	–	De-Green Fos-Fall-A	defoliant desiccant	–	6, 20
tributylphosphorotrithioite	merphos	–	Folex Easy-Off-D	defoliant	–	6
trichloroacetic acid	TCA	–	Varitox NaTA	desiccant	herbicide	20
N-trichloroacetylaminomethylenephosphonic acid	–	–	–	sugarcane ripener	–	17
2,3,6-trichlorobenzoic acid, dimethylamine salt	2,3,6-TBA	TBA	Trysben	sugarcane ripener	herbicide	17
2,4,5-trichlorophenoxy acetic acid	2,4,5-T	–	Fruitone A Weedone Weedar Brush Killer	root inducer latex stimulator fruit ripener	herbicide	2, 17
2,4,6-trichlorophenoxy acetic acid	–	–	–	herbicide antidote	–	21
2-(2,4,5-trichlorophenoxy) propanoic acid	2,4,5-TP	–	Fruitone-T	fruit drop preventer fruit ripener	–	17
bis(N,O-trifluoroacetyl)-N-phosphonomethyl-glycine	–	–	–	sugarcane ripener	–	17
3-trifluoromethyl-sulfonamido-p-acetotoluidide	fluoridamid	MBR-6033	Sustar	growth retardant sugarcane ripener	herbicide	8, 17

Table 24–1. (continued)

Chemical Name	Common Name	Designation or Code	Trade Name	PGR Activity	Other Activity	Chapter Location
2,4a,7-trihydroxy-1-methyl-8-methylene-gibb-3-ene-1,10-carboxylic acid-1,4-lactone	gibberellic acid	GA_3	Pro-Gibb Gro-Sol Gib-Tabs Gibrel	senescence delayer sex changer germination enhancer amylase stimulant flowering inducer shoot growth stimulant seedless grape enhancer flowering delayer fruit set enhancer gametocide insect controller disease controller ripening delayer	–	3, 4, 5, 7, 8, 13, 15, 17, 18, 19
2,3,5-triiodobenzoic acid	TIBA	–	Regime 8	soybean stimulator gametocide tillering agent disease controller sex changer	–	5, 12, 13, 17, 18

Literature

1. Nickell LG (1978) Chem Eng News *56* (41), 18
2. Thimann KV (1974) Plant Physiol *54,* 450
3. Went FW (1926) Proc Kon Ned Akad Wetensch *C30,* 10
4. Kögl F, Haagen-Smit AJ, Erxleben H (1934) Z physiol Chem *228,* 90
5. Rodriguez AB (1932) Dept Agri Puerto Rico *16,* 1
6. Hitchcock AE, Zimmerman PW (1940) Contrib Boyce Thompson Inst *10,* 461
7. Hitchcock AE, Zimmerman PW (1940) Contrib Boyce Thompson Inst *11,* 143
8. Mitchell JW (1966) Agri Sci Rev *4* (4), 27
9. Nickell LG (1981) Encyclopedia of Chemical Technology, 3rd edn, vol 17, in press
10. Avery GS Jr., Johnson EF (1947) Hormones and Horticulture, New York, McGraw Hill
11. Hartmann HT, Kester DE (1968) Plant propagation: Principles and practices, Englewood Cliffs, N. J., Prentice Hall
12. Leopold AC (1955) Auxins and plant growth, Berkeley and Los Angeles, Univ. of Calif. Press
13. Thimann KV, Behnke-Rogers J (1950) Maria Moors Cabot Found Publ No 1, 1
14. Weaver RJ (1972) Plant growth substances in agriculture, San Francisco, Freeman
15. Audus LJ (1953) Plant growth substances, London, Leonard Hill Ltd.
16. Khosh-Khui M, Tafazoli E (1979) Sci Hort *10,* 395
17. Das P, Mahapatra P, Das RC (1978) Orissa Jour Hort *6,* 31
18. Rashid A, Badrul Alam AFM (1976) Tea Jour Bangladesh *12,* 13
19. Lawhead CW, Bennett JP, Yamaguchi M (1979) Trop Agr *56,* 271
20. Hieke K (1979) Deutsche Baumschule *31,* 376
21. Ylatalo M (1979) Jour Sci Agri Soc Finland *51,* 163
22. Alley CJ (1980) Calif Agri *34* (7), 29
23. Mokashi AN (1978) Mysore Jour Agri Sci *12,* 528
24. Sympson RL, Chin C (1980) HortScience *15,* 309
25. Rousseau GG (1965) So Afr Jour Agri Sci *8,* 1173
26. Rousseau GG (1967) Farming So Afr *42* (12), 58
27. Criley RA, Parvin PE (1979) Jour Amer Soc Hort Sci *104,* 592
28. Davies FT Jr, Joiner JN (1980) Jour Amer Soc Hort Sci *105,* 91
29. Rauch FD, Yamakawa RM (1980) HortScience *15,* 97
30. Mukherjee TP, Roy T, Bose TK (1976) Punjab Hort Jour *16,* 153
31. Singh SP (1979) Prog Hort *11,* 49
32. Siddiqui BA, Khan IA (1978) Acta Bot Ind 6, 128
33. Kramer PJ, Kozlowski TT (1960) Physiology of Trees. New York. McGraw-Hill
34. Gorecki RS (1979) Acta Agrobot 32, 223
35. Bartolini G, Bellini E, Messeri C (1979) Riv Ortoflorofrutticol 63, 423
36. Nemeth G (1979) Z Pflanzenphysiol *95,* 389
37. Tarasenko MT, Balobin VN, Fedurko T (1979) Izvestiya Timiryazevskoi Sel'skokhozyaistvennoi Akademii *6,* 104
38. Fadl MS, El-Deen SAS, El-Mahdy A (1979) Egypt Jour Hort *6,* 55
39. Lahiri AK (1979) Ind For *105,* 101
40. Kiang TY, Rogers OM, Dike RB (1977) NZ Jour For Sci *4,* 153
41. Thomas JK, Riker AJ (1950) Jour For *48,* 474
42. Ja'afar H, Pakaniathan SW (1979) Jour Rubber Res Inst Malaysia *27,* 143

43. Isikawa H, Tanaka I (1970) Jour Pop Soc *52,* 99
44. Florido LV (1978) Sylvatrop Philipp For Res Jour *3,* 115
45. Florido LV, Limsuan MP (1977) Sylvatrop Philipp For Res Jour *2,* 55
46. Enright LJ (1957) Jour For *55,* 892
47. Hartung W, Ohl B, Kummer V (1980) Z Pflanzenphysiol *98,* 95
48. Chin TY, Meyer M, Beevers L (1969) Planta *88,* 192
49. Blazich FA, Heuser CW (1977) HortScience *12,* 392
50. Rasmussen S, Andersen AS (1980) Z Pflanzenphysiol *98,* 95
51. Basu RN, Roy BN, Bose TK (1970) Plant Cell Physiol *11,* 681
52. Juillard B (1970) CR Acad Sci (Paris) *D270,* 2795
53. Schmid A (1972) Ber Schweiz Bot Ges *82,* 14
54. Read EP, Hoysler V (1971) HortScience *6,* 350
55. Krishnamoorthy HN (1972) Biochem Physiol Pflanzen *163,* 513
56. Krelle E, Libbert E (1969) Flora *160,* 299
57. Coleman W, Greyson RI (1976) Jour Exp Bot *27,* 1339
58. Delegher-Langohr V (1974) Bull Soc Roy Bot Belg *107,* 73
59. Biddington NL (1978) Brit Plant Growth Regulator Group Monogr *2,* 29
60. Zigas RP, Coombe BG (1977) Aust Jour Plant Physiol *4,* 359
61. Hsiao AI (1980) Can Jour Plant Sci *60,* 643
62. Hsiao AI (1979) Can Jour Bot *57,* 1729
63. Abdalla KM, El-Wakeel AT, El-Masiry HH (1978) Res Bull Ain Shams Univ Fac Agri *944,* 1
64. Burns RM, Coggins CW (1969) Calif Agri *23,* 18
65. Palevitch D, Thomas TH (1976) Physiol Plant *37,* 247
66. Thomas TH, Palevitch D, Austin RB (1972) Proc. 11th Brit. Weed Control Conf., p 760
67. Palevitch D, Thomas TH (1975) Physiol Plant *34,* 134
68. Yen ST, Carter OG (1972) Aust Jour Exp Agri Animal Husbandry *12,* 653
69. Coats GE (1967) Miss Agr Exp Stn Bull *752,* 1
70. Coal DF, Wheeler JE (1974) Crop Sci *14,* 451
71. Wittwer SH, Bukovac MJ (1957) Mich Agri Exp Stn Quart Bull *40,* 215
72. Kojima H, Oota Y (1980) Plant Cell Physiol *21,* 561
73. Villareal RL (1980) Linking Research to Crop Production. In: Staples RC, Kuhr RJ (eds) New York, Plenum Press, p 1
74. Nelson JM, Sharples GC (1980) HortScience *15,* 253
75. Renard HA, Clerc P (1978) Seed Sci Technol *6,* 661
76. Ikuma H, Thimann KV (1960) Plant Physiol *35,* 557
77. Lona F (1956) L'ateneo Parmanse *27,* 641
78. Miller CO (1956) Plant Physiol *31,* 318
79. Poljakoff-Mayber A (1958) Bull Res Council Israel *60,* 78
80. Mayer AM, Poljakoff-Mayber A (1963) The germination of seeds, New York, Pergamon Press
81. Khan AA, Tolbert NE (1966) Physiol Plant *19,* 81
82. Evenari M, Mayer AM (1954) Bull Res Council Israel *4,* 81
83. Berrie AMM, Robertson J (1973) Physiol Plant *28,* 278
84. Lado P, Rasi-Caldogno F, Colombo R (1974) Physiol Plant *31,* 149
85. Halloin JM (1976) Plant Physiol *57,* 454
86. Amen RD (1968) Bot Gaz *34,* 1
87. Samish RM (1954) Ann Rev Plant Physiol *5,* 183
88. Weaver RJ (1972) Plant growth substances in agriculture. San Francisco, W. H. Freeman
89. Kennedy EJ, Smith O (1953) Proc Amer Soc Hort Sci *61,* 395
90. Marshall ER, Smith O (1951) Bot Gaz *112,* 329
91. Sawyer RL, Dallyn SL (1958) Amer Potato Jour *35,* 620
92. Wittwer SH, Sharma RC (1950) Science *112,* 597
93. Rappaport L (1956) Calif Agri *10,* 4
94. Rappaport L, Lippert LF, Timm H (1957) Amer Potato Jour *34,* 254
95. Timm H, Rappaport L, Bishop JC, Hoyle BJ (1962) Amer Potato Jour *39,* 107

96. Rodriguez AB (1932) Jour Dept Agri Puerto Rico *26,* 5
97. Gowing DP, Leeper RW (1953) Science *122,* 1267
98. Abeles FB (1973) Ethylene in Plant Biology. New York, Academic Press
99. Cooke AR, Randall DI (1968) Nature *218,* 974
100. Seeyave J (1966) Rev Agri Sucre Ile Maurice *45,* 299
101. Guyot A, Py C (1970) Fruits d'Outre Mer *25,* 427
102. Clark HE, Kearns KR (1942) Science *95,* 536
103. Cooper WC (1942) Proc Amer Soc Hort Sci *41,* 93
104. Cibes HR, Gandia H (1962) Jour Agri Univ Puerto Rico *46,* 65
105. Singmaster JA (1970) Jour Agri Univ Puerto Rico *54,* 184
106. Evans HR (1959) Trop Agri (Trinidad) *36,* 108
107. Das N (1964) Indian Jour Agri Sci *34,* 38
108. Wee YC, Ng JC (1971) Malaysian Pineapple *1,* 5
109. Nyenhuis EN (1968) Farming So Afr *43* (9), 31
110. Wu WL (1966) Taiwan Agri Quart *2* (2), 42
111. Apte SS (1969) Ghana Jour Agri Sci *2,* 48
112. Norman JC (1972) Ghana Jour Agri Sci *5,* 213
113. Norman JC (1977) Sci Hort *7,* 143
114. Luckwill LC (1975) Proc. XIX Internat'l. Hort. Cong. (Warsaw) 3, 235
115. Edgerton LJ (1966) Proc Amer Soc Hort Sci *88,* 197
116. Sitton BG, Lewis WA, Kilby WW (1968) Proc Amer Soc Hort Sci *92,* 381
117. Proebsting EL, Mills HH (1974) Jour Amer Soc Hort Sci *99,* 464
118. Weaver RJ (1959) Nature *183,* 1198
119. Brian PW, Petty JHP, Richmond PT (1959) Nature *184,* 69
120. Corgan JN, Widmoyer FB (1971) Jour Amer Soc Hort Sci *96,* 54
121. Proebsting EL, JR., Mills HH (1964) Jour Amer Soc Hort Sci *85,* 134
122. Stembridge GE, LaRue JH (1969) Jour Amer Soc Hort Sci *94,* 492
123. Dennis FG (1976) Jour Amer Soc Hort Sci *101,* 241
124. Hicks JR, Crane JC (1968) Proc Amer Soc Hort Sci *92,* 1
125. Griggs W, Iwakiri G, Bethell R (1965) Calif Agri *19,* 9
126. Sullivan DT, Widmoyer FB (1970) HortScience *5,* 91
127. Gil G, Bruzzone E (1972) Agri Tecn (Chile) *33* (1), 30
128. Gil G, Gomez E, Ryugo K (1972) Agri Tecn (Chile) *32* (2), 79
129. Gil G, Yrarrazaval F (1974) Ciencia e Invest Agraria *1,* 163
130. Garner WW, Allard HA (1920) Jour Agri Res *18,* 553
131. Nickell LG (1976) Outlook on Agri *9,* 57
132. Nickell LG (1980) Plant Growth Substances 1979. In: Skoog F (ed) Berlin-Heidelberg-New York, Springer, p 419
133. Tanimoto TT, Nickell LG (1965) Proc. 12th Internat'l. Soc. Sugar Cane Technol., p 113
134. Humbert RP, Lima M, Goveas J (1968) Proc. 13th Internat'l. Soc. Sugar Cane Technol., p 462
135. Benedicto F (1967) Victorias Milling Co. Exp. Stn. Bull, March-April, p 3
136. Yang PC, Pao TP, Ho FW (1972) Taiwan Sugar *19* (1), 21
137. Thomas RO (1975) Crop Sci *15,* 87
138. Cathey GW, Thomas RO (1979) Proc. 33rd Beltwide Cotton Prod. Res. Conf., p 277
139. Hacskaylo J, Scales AL (1959) Jour Econ Entomol *52,* 396
140. Brown LC, Lincoln C, Frans RF, Waddle BA (1961) Jour Econ Entomol *54,* 309
141. Brown LC, Cathey GW, Lincoln C (1962) Jour Econ Entomol *55,* 298
142. Roark B, Pfrimmer TR, Merkle ME (1963) Crop Sci *3,* 338
143. Phillips JR, Herzog GA, Nicholson WF (1977) Ark Farm Res *27,* 4
144. Stuart NW (1961) Science *134,* 50
145. Stuart NW (1962) Florists Rev *194,* 35
146. Marth PC (1963) Proc Amer Soc Hort Sci *83,* 777
147. Deidda P (1972) Riv Ortoflorofrutticoltura Italiana *5–6,* 888
148. Deidda P, Agabbio M (1977) Proc Internat'l Soc Citriculture *2,* 688
149. Moss GI (1973) First Cong Mund Citricultura, Murcia, Valencia (Spain) *2,* 215
150. Moss GI (1973) First Cong Mund Citricultura, Murcia, Valencia (Spain) *2,* 367

151. Monselise SP, Halevy AH (1964) Proc Amer Soc Hort Sci *84,* 141
152. Monselise SP, Goren R, Halevy AH (1966) Proc Amer Soc Hort Sci *89,* 195
153. Moss GI (1971) Aust Jour Agri Res *22,* 625
154. El-Shafie SA (1978) Arch Gartenbau *26,* 287
155. El-Motaz-Bellah M, Erafa AE, Shahin H (1977) Agr Res Rev (Egypt) *55,* 211
156. Zieslin N, Madori G, Halevy AH (1979) Jour Exp Bot *30,* 15
157. Bottle RT (1964) Manuf Chem *35,* 60
158. Valio IFM, Rocha RF (1977) Japan Jour Crop Sci *46,* 243
159. Jones RL (1973) Ann Rev Plant Physiol *24,* 571
160. Alamu S, McDavid CR (1978) Trop Agri *55,* 81
161. Harbaugh BK, Wilfret GJ (1979) HortScience *14,* 72
162. Henry RJ (1980) HortScience *15,* 613
163. McDavid CR, Alamu S (1976) Trop Agri *54,* 373
164. McDavid CR, Alamu S (1979) Trop Agri *56,* 17
165. Fisher JB (1980) Jour Exp Bot *31,* 731
166. Pharis RP, Kuo CG (1977) Can Jour For Res *7,* 299
167. Dunberg A (1974) Stud For Suecica *111,* 1
168. Pharis RP, Ross SD, McMullan E (1980) Physiol Plant *50,* 119
169. Hashizume H (1967) Jour Jap For Soc *49,* 405
170. Ross SD (1976) Acta Hort *56,* 163
171. Ross SD, Pharis RP (1976) Physiol Plant *36,* 182
172. Puritch GS, McMullan EE, Meagher MD, Simmons CS (1979) Can Jour For Res *9,* 193
173. Tompsett PB (1977) Ann Bot *41,* 1171
174. Tompsett PB (1978) Proc. Joint BCPC and PGRG Symposium "Opportunities for Chemical Plant Growth Regulation", p 75
175. Tompsett PB, Fletcher AM (1979) Physiol Plant *45,* 112
176. Tompsett PB, Fletcher AM, Arnold GM (1980) Ann Appl Biol *94,* 421
177. Ross SD, Greenwood MS (1979) Physiol Plant *45,* 207
178. Batch JJ (1978) Proc Joint BCPC and BPGRG Symp. "Opportunities for Chemical Plant Growth Regulation" p 33
179. Naylor AW (1950) Proc Nat'l Acad Sci (USA) *36,* 230
180. Moore RH (1950) Science *112,* 52
181. Rehm S (1952) Nature *170,* 38
182. Fairey DT, Stoskopf NC (1975) Crop Sci *15,* 29
183. McNulty PJ, Warner HL: US 4 028 084 (11.10.74/7.6.77)
184. Brown CM, Earley EB (1973) Agron Jour *65,* 829
185. Hughes WG, Bennett MD, Bodden JJ, Galanopoulou S (1974) Ann Appl Biol *76,* 243
186. Rowell PL, Miller DG (1971) Crop Sci *11,* 629
187. Rowell PL, Miller DG (1974) Crop Sci *14,* 31
188. Stoskopf NC, Law J (1972) Can Jour Plant Sci *52,*, 680
189. Varenitsa ET, Sotnik VM (1973) Dokl Vses Akad Sel'skokhoz Nauk *4,* 3
190. Hockett EA (1972) Barley Newsltr *15,* 95
191. Law J, Stoskopf NC (1973) Can Jour Plant Sci *53,* 765
192. Sapra VT, Sharma GC, Hughes JL (1974) Euphytica *23,* 685
193. Hecker RJ, Bilgen T, Bhatnagar PS, Smith GA (1972) Can Jour Plant Sci *52,* 937
194. Hecker RJ, Smith GA (1975) Can Jour Plant Sci *55,* 655
195. Lower RL, Miller CH (1969) Nature *222,* 1072
196. McMurray AL, Miller CH (1969) Jour Amer Soc Hort Sci *94,* 400
197. Robinson RW, Whitacker TW, Bahn GW (1970) Euphytica *19,* 180
198. Jan CC, Qualset CO, Vogt HE (1974) Euphytica *23,* 78
199. Trupp CR (1972) Amer Soc Agron Absts, p 29
200. Wang RC, Lund S (1975) Crop Sci *15,* 550
201. Laible CA (1974) Proc. 29th Ann. Corn & Sorghum Res. Conf., p 174
202. Theurer JC (1979) Can Jour Plant Sci *59,* 463
203. Johnson RR, Brown CM (1976) Crop Science *16,* 584
204. Starke GR, Cooke AR: US 4 009 020 (6.5.75/22.2.77)

205. Fedin MA, Gyska MN, Penyazkova NV, Sochilin EC, Al'sing TK: USSR 640 710 (1.9.75/5.1.79)
206. Miller DA, Hittle CN (1963) Crop Sci *3,* 397
207. Choudhurry AR (1966) Pakistan Jour Sci *18,* 175
208. Eaton FW (1957) Science *126,* 1175
209. Gagneja MR, Singh TP (1973) Ind Jour Gen Plant Breeding *33,* 340
210. Natrova Z, Hlavic M (1975) Biol Plant *17,* 251
211. Foster CA (1969) Ann Bot *33,* 947
212. de Sacks RL, Rosas GS (1966) Plant Breeding Absts *33,* 1654
213. Dudley JW (1960) Jour Amer Soc Sugar Beet Technol *11,* 25
214. Moore JF (1959) Science *129,* 1738
215. Moore JF (1964) Proc Amer Soc Hort Sci *84,* 474
216. Porter KB, Wiese AF (1961) Crop Sci *1,* 381
217. Arumagan R, Rao VHM (1973) Ind Jour Plant Physiol *16,* 1
218. Chauhan SVS, Singh SP (1972) Ind Jour Plant Physiol *15,* 138
219. Chopra VL, Jain SK, Swaminanthan MS (1960) Ind Jour Gen Plant Breeding *20,* 188
220. Bollinger FJ, D'Amico JJ, Hansen DJ: US 4 124 375 (12.10.77/7.11.78)
221. Fedin MA, Gyska MN, Penyazkova NV, Sochilin EG, Al'sing TK: USSR 640 711 (1.9.75/5.1.79)
222. Johnson WO, Seidel MC, Warner HL: US 4 038 065 (11.10.74/26.6.77)
223. Fedin MA, Gyska MN, Penyazkova NV, Terent'ev AG, Ikonnikow NS: USSR 640 712 (20.10.75/5.1.79)
224. Van der Meer OP, Van Dam R (1979) Euphytica *28,* 717
225. Hansen DJ, Bellman SK, Sacher RM (1975) Plant Physiol *56* (Suppl), 44
226. Nelson PM, Rossman EC (1958) Science *127,* 1500
227. Wittwer SH, Bukovac MJ (1958) Econ Bot *12,* 213
228. Eenink AH, Loupias S (1976) Zaadbelangen *10,* 300
229. Van der Meer OP, Van Bennekom JL (1973) Euphytica *22,* 239
230. Soethara A, Kumari PK (1975) Ind Jour Gen Pl Breeding *35,* 136
231. Cantliffe DJ (1974) Can Jour Pl Sci *54,* 771
232. Kaushik MP, Sharma JK (1974) Ind Jour Exp Biol *12,* 599
233. Jan CC, Qualset CO, Vogt HE (1976) Euphytica *25,* 375
234. Varenitsa ET, Popov BV (1980) Vestn S-kh Nauk (Moscow) *1,* 21
235. Cathey GW (1979) Outlook Agriculture *10,* 191
236. Crowe GB, Carns HR (1957) Miss Agri Exp Stn Bull *552,* 1
237. Brendel TP, Miller CS (1979) Proc. Beltwide Cotton Prod. Res. Conf. *33,* 56
238. Miller CS, Brendel TP (1979) Proc. Beltwide Cotton Prod. Res. Conf. *33,* 56
239. Arendt F, Rusch R, von Stillfried H, Hanisch B, Martin WC (1976) Plant Physiol *57* (Suppl), 99
240. Kittock DL, Arle HF, Bariola AL (1975) Proc. Beltwide Cotton Prod. Res. Conf. *29,* 71
241. Cathey GW (1978) Crop Sci *18,* 301
242. Wilson WC (1973) Acta Hort *34,* 377
243. Hendershott CH (1965) Proc Fla State Hort Soc *78,* 36
244. Wilson WC (1967) Proc Fla State Hort Soc *80,* 277
245. Wilson WC, Coppock GE (1968) Proc Fla State Hort Soc *81,* 39
246. Wilson WC (1969) Proc Fla State Hort Soc *82,* 75
247. Wilson WC (1971) Proc Fla State Hort Soc *84,* 67
248. Wilson WC (1973) Proc Fla State Hort Soc *86,* 56
249. Wilson WC (1973) HortScience *8,* 323
250. Kenney DS, Clark RK, Wilson WC (1974) Proc Fla State Hort Soc *87,* 34
251. Clark RK, Wilson WC (1975) Proc Fla State Hort Sci *88,* 100
252. Holm RE, Wilson WC (1975) Proc Fla State Hort Soc *88,* 103
253. Davies FS, Cooper WC, Galena FE (1975) Proc Fla State Hort Soc *88,* 107
254. Clark RK, Ellis MT (1976) Proc Fla State Hort Soc *89,* 72
255. Freeman B, Sarooshi RA (1976) Aust Jour Exp Agri Animal Husbandry *16,* 943
256. Rasmussen GK (1975) HortScience *10,* 516
257. Wilson WC, Holm RE (1976) Proc Fla State Hort Soc *89,* 32

258. Holm RE, Wilson WC (1977) Jour Amer Soc Hort Sci *102,* 576
259. Wilcox M, Taylor JB, Wilson WC, Chen IY (1974) Proc Fla State Hort Soc *87,* 22
260. Wilcox M, Chen IY, Taylor JB, Wilson WC, Li YY (1977) Proc. Plant Growth Regulator Working Group *4,* 246
261. Lavee S, Barshi G, Haskal A (1973) Sci Hort *1,* 63
262. Vitagliano C (1975) Jour Amer Soc Hort Sci *100,* 482
263. Ben-Tal Y, Lavee S (1976) HortScience *11,* 489
264. Troncoso A, Prieto J, Liñan J (1978) An Edafol Agrobiol *37,* 881
265. Hartmann HT, Fadl M, Whisler J (1967) Calif Agri *21,* 5
266. Hartmann HT, Reed W, Opitz K (1976) Jour Amer Soc Hort Sci *101,* 278
267. Hartmann HT (1955) Bot Gaz *117,* 24
268. Hartmann HT, Heslop AJ, Whisler J (1968) Calif Agri *22,* 14
269. Hartmann HT, El-Hamady M, Whisler J (1972) Jour Amer Soc Hort Sci *97,* 781
270. Vitagliano C (1969) Sci Tec Agri *2,* 160
271. Hartmann HT, Tombesi A, Whisler J (1970) Jour Amer Soc Hort Sci *95,* 535
272. Lavee S, Haskal A (1975) Sci Hort *3,* 163
273. Donno G, Ferrara E, Reina A (1974) Ann Fac Agraria Univ Bari *27,* 197
274. Vitagliano C, Zucconi R (1973) Sci Tec Agri *6,* 185
275. Rufener J, della Pieta S (1974) Riv Ortoflorofruttic Ital *44,* 156
276. Martin GC, Lavee S, Sibbett GS, Nishijima C, Carlson SP (1980) Calif Agr *34,* 7
277. Williams MW (1979) Hort Rev *1,* 270
278. Auchter EC, Roberts JW (1934) Proc Amer Soc Hort Sci *30,* 23
279. Magness JR, Batjer LP, Harley CP (1940) Proc Amer Soc Hort Sci *37,* 141
280. Batjer LP, Thompson AH (1948) Proc Amer Soc Hort Sci *52,* 164
281. Burkholder CL, McCown M (1941) Proc Amer Soc Hort Sci *38,* 117
282. Schneider GW, Enzie JV (1943) Proc Amer Soc Hort Sci *42,* 167
283. Davidson JH, Hammer OH, Reimer CA, Dutton WC (1945) Mich Agri Exp Stn Quart Bull *27,* 352
284. Batjer LP, Westwood MN (1960) Proc Amer Soc Hort Sci *75,* 1
285. Bukovac MJ, Mitchell AE (1961) Proc Amer Soc Hort Sci *80,* 1
286. Stebbins RL (1962) Proc Amer Soc Hort Sci *80,* 11
287. Horsfall F, Moore RC (1963) Proc Amer Soc Hort Sci *82,* 1
288. Williams MW, Batjer LP (1964) Proc Amer Soc Hort Sci *85,* 1
289. Robitaille HA, Emerson FH, Yu KS (1977) Jour Amer Soc Hort Sci *102,* 595
290. Byers RE (1978) HortScience *13,* 59
291. Chiba K, Kubota T (1979) Bull Fruit Tree Res Stn (Morioka) *6,* 55
292. Kabluchko GA, Limarenko AM (1979) Sadovodstvo *7,* 18
293. HerraraAguirre E, Unrath CR (1980) HortScience *15,* 43
294. Edgerton LJ, Greenhalgh WJ (1969) Jour Amer Soc Hort Sci *94,* 11
295. Veinbrants N, Hutchinson JF (1976) Aust Jour Exp Agri Animal Husbandry *16,* 937
296. Byers RE (1978) Jour Amer Soc Hort Sci *103,* 232
297. Zucconi F (1978) Acta Hort *80,* 245
298. Stembridge GE, Gambrell CE (1969) Jour Amer Soc Hort Sci *94,* 570
299. Gerin G, Giulivo C (1972) Riv Ortoflorofruttic Ital *5–6,* 699
300. Morini S, Vitagliano C, Xiloyannis C (1972) Riv Ortoflorofruttic Ital *5–6,* 615
301. Gerin G (1973) L'Informatore Agrario *48,* 14351
302. Morini S, Vitagliano C, Xiloyannis C (1976) Jour Amer Soc Hort Sci *101,* 640
303. Daniell JW (1975) HortScience *13,* 345
304. Byers RE (1976) HortScience *11,* 324
305. Gambrell CE, Stembridge GE (1978) HortScience *13,* 261
306. Porpiglia PJ, Barden JA (1980) Jour Amer Soc Hort Sci *105,* 227
307. Thompson AH, Rogers BL (1959) Proc Amer Soc Hort Sci *73,* 112
308. Martin GC, Nelson MM, Nishijima C (1971) HortScience *6,* 169
309. Daniell JW, Wilkinson RE (1972) Jour Amer Soc Hort Sci *97,* 682
310. Thompson AH, Rogers BL (1972) Jour Amer Soc Hort Sci *97,* 644
311. Antognozzi E (1972) Riv Ortoflorofruttic Ital *5–6,* 634
312. Weinbaum SA, Giulivo C, Ramina A (1977) Jour Amer Soc Hort Sci *102,* 781

313. Keil HL, Fogle HW (1971) HortScience *6,* 403
314. Marth PC, Prince VE (1953) Science *117,* 497
315. Horsfall F, Moore RC (1956) Proc Amer Soc Hort Sci *68,* 63
316. Stembridge GE, Gambrell CE (1971) Jour Amer Soc Hort Sci *96,* 7
317. Young E, Edgerton LJ (1979) HortScience *14,* 173
318. Rom RC, Arrington EH, Brown SA (1980) Ark Farm Res *29* (3), 12
319. Dhuria HA, Bhutani VP, Parmar C (1976) Sci Hort *4,* 279
320. Kvale A (1978) Acta Agri Scand *28,* 279
321. Geiszler J, Mady R, Toth M (1974) Kert Egy Közl *38,* 195
322. Toth M (1976) Kert Egy Közl *40,* 139
323. Geiszler J, Mady R, Toth M (1975) Kertgazdasag *7,* 15
324. Szabo Z, David M, Toth M (1978) Acta Agron Acad Sci Hung *27,* 405
325. Micke WC, Schreader WR, Yeager JT, Roncoroni EJ (1975) Calif Agri *29* (8), 3
326. Lal BB, Thakur DR (1978) Haryana Jour Hort Sci *7,* 21
327. Gallasch PT (1974) Aust Jour Exp Agri Animal Husbandry *14,* 835
328. Gallasch PT (1978) Aust Jour Exp Agri Animal Husbandry *18,* 152
329. Iwahori S, Oohata JT (1976) Sci Hort *4,* 167
330. Hield HZ, Burns RM, Coggins CW (1962) Proc Amer Soc Hort Sci *81,* 218
331. Hirose K, Iwagaki I, Suzuki K (1978) Absts 20th Internat Hort Cong *11*
332. O'Mara RM (1977) West Aust Nutgrowing Soc Yearbook *3,* 29
333. Olson WH, Sibbett GS, Carnill GL, Martin GC (1977) Calif Agri *31* (7), 6
334. Miele A, Weaver RJ, Johnson JO (1978) Vitis *17,* 369
335. Weaver RJ (1954) Proc Amer Soc Hort Sci *63,* 194
336. Samish RM, Lavee S (1958) Ktavim Rec Agri Res Stn *8,* 273
337. Weaver RJ (1963) Vitis *4,* 1
338. Zuluaga EM, Lunelli J, Christensen JH (1968) Phyton *25,* 35
339. Eynard I (1970) Proc. 10th Br. Weed Control Conf. 275
340. Hull J, Bukovac MJ, Howell GS (1970) HortScience *5,* 348
341. Peterson JR, Hedberg PR (1975) Sci Hort *3,* 275
342. Clore WJ, Fay RD (1978) HortScience *5,* 21
343. Hedberg PR, Goodwin PB (1980) Amer Jour Enol Vitic *31,* 109
344. Browning G, Cannell MGR (1970) Jour Hort Sci *45,* 223
345. Adenikinju SA (1975) Turrialba *25,* 414
346. Yamamura H, Naito R (1975) Jour Jap Soc Hort Sci *43,* 406
347. Yamamura H, Naito R, Mochida K (1976) Jour Jap Soc Hort Sci *45,* 1
348. Yamamura H, Naito R (1980) Jour Jap Soc Hort Sci *49,* 171
349. Krezdorn AH (1969) Proc 1st Internat'l Citrus Symp *3,* 1113
350. Krezdorn AH, Jernberg DC (1977) Proc Internat'l Soc Citriculture, *2,* 660
351. Weaver RJ (1952) Proc Amer Soc Hort Sci *60,* 132
352. Weaver RJ (1955) Proc Amer Soc Hort Sci *65,* 183
353. Weaver RJ (1973) Acta Hort *34,* 275
354. Looney NE (1975) Can Jour Plant Sci *55,* 117
355. Tukey LD, Flemming HK (1968) Proc Amer Soc Hort Sci *93,* 300
356. Coombe BG (1967) Vitis *6,* 278
357. Barritt BH (1970) Jour Amer Soc Hort Sci *95,* 58
358. Tafazoli E (1977) Sci Hort *6,* 121
359. Weaver RJ, van Overbeek J, Pool RM (1966) Hilgardia *37,* 181
360. Weaver RJ, van Overbeek J (1963) Calif Agri *17* (9), 12
361. McCollum JP (1934) Cornell Agri Exp Stn Mem *163,* 1
362. Cantliffe DJ (1972) Can Jour Plant Sci *52,* 781
363. Cantliffe DJ, Robinson RW, Bastdorff RS (1972) HortScience *7,* 285
364. Cantliffe DJ, Robinson RW, Shannon S (1972) HortScience *7,* 416
365. Robinson RW, Shannon S, de la Guardia M (1969) BioScience *19,* 141
366. Cantliffe DJ (1976) Proc Fla State Hort Soc *89,* 94
367. Olympios CM (1976) Hort Res *16,* 65
368. Murthy KN, Kumaran PM, Nayar NM (1975) Jour Plantation Crops *3,* 81
369. Dua IS, Bhardwaj SN (1979) Ind Jour Plant Physiol *12,* 50

370. Stahly EA, Williams MW (1976) HortScience *11,* 502
371. Bangerth F (1978) Jour Amer Soc Hort Sci *103,* 401
372. Wang CY, Mellenthin WM (1977) Plant Physiol *59,* 546
373. Williams MW (1980) HortScience *15,* 76
374. Nyeki J, Soltesz M, Tisza A (1977) Vjabb kutatasi eredmenyek a gyümölestermesztesben (Hungary), p 89–109
375. Rojas Garciduenas M, Bustamante M, Siller A (1971) Turrialba *21,* 169
376. Lukasik S (1975) Acta Agrobot *28,* 43
377. de Silva WH, Bocion PF, Eggenberg P, deMur A (1979) Ztschr Pflanzenkrankheiten Pflanzenschutz *86,* 546
378. Ferre DC, Stang EJ (1980) Ohio Agri Res Dev Res Circ *259,* 4
379. Cibulsky RJ, Greene GM (1979) Proc Plant Growth Regulator Working Group *6,* 180
380. Greenhalgh WJJ, Godley GL (1976) Aust Jour Exp Agri Animal Husbandry *16,* 592
381. Greenhalgh WJJ, Godley GL, Meuzies R (1977) Aust Jour Exp Agri Animal Husbandry *17,* 505
382. Yabuta T, Sumiki Y (1938) Jour Agri Chem Soc (Japan) *14,* 1526
383. Tanimoto TT, Nickell LG (1966) Rpts. 26th Ann. Conf. Haw. Sugar Technol., p 184
384. Tanimoto TT, Nickell LG (1967) Rpts. 27th Ann. Conf. Haw. Sugar Technol., p 137
385. Moore PH (1978) Proc Plant Growth Regulator Working Group *5,* 158
386. Moore PH, Ginoza H (1980) Crop Sci *20,* 78
387. Ries SK, Wert V, Sweeley CC, Leavitt RA (1977) Science *195,* 1339
388. Ries SK, Sweeley CC: US 4 150 970 (3.1.77/24.4.79)
389. Ries SK, Richman TL, Wert VF (1978) Jour Amer Soc Hort Sci *103,* 361
390. Ries SK, Richman TL, Wert VF (1978) Proc Plant Growth Regulator Working Group *5,* 114
391. Ries SK, Wert VF (1977) Planta *135,* 77
392. Hangarter R, Ries SK, Carlson P (1978) Plant Physiol *61,* 855
393. Bittenbender HC, Dilley DR, Wert VF, Ries SK (1978) Plant Physiol *61,* 851
394. Singletary GW, Foy CL (1980) Proc Plant Growth Regulator Working Group *7,* 29
395. Bouwkamp JC, McArdle RN (1980) HortScience *15,* 69
396. Charlton JL, Hunter NR, Green NA, Fritz W, Addison BM, Woodbury W (1980) Can Jour Plant Sci *60,* 795
397. Marcelle RD, Chrominski A (1978) Proc Plant Growth Regulator Working Group *5,* 116
398. Chowdhury IR, Paul KB, Sasseville D (1980) Proc Plant Growth Regulator Working Group *7,* 76
399. Steffens GL, Worley JF (1980) Proc Plant Growth Regulator Working Group *7,* 137
400. Ohlrogge AJ, Fulk-Bringman SS (1980) Proc Plant Growth Regulator Working Group *7,* 138
401. Sagaral EG, Orcutt DM, Foy CL (1978) Proc Plant Growth Regulator Working Group *5,* 115
402. Jones J, Wert V, Ries S (1979) Plant *144,* 277
403. Ries SK (1979) Proc Plant Growth Regulator Working Group *6,* 92
404. Ogawa M, Matsui T, Tobitsuka J (1978) Phytochem *17,* 343
405. Ogawa M, Kitamura H (1980) Planta *147,* 495
406. Ogawa M, Matsui T, Oyamada K, Tobitsuka J (1977) Plant Cell Physiol *18,* 841
407. Grove MD, Spencer GF, Rohwedder WK, Mandava N, Worley JF, Warthen JD, Steffens GL, Flippen-Anderson JL, Cook JC (1979) Nature *281,* 216
408. Steffens GL (1979) Proc Plant Growth Regulator Working Group *6,* 86
409. Mandava N, Kozempel M, Worley JF, Matthees D, Warthen JD, Jacobson M, Steffens GL, Kenney H, Grove MD (1978) Ind Eng Chem Prod Res Dev *17,* 351
410. Thompson MJ, Mandava N, Flippen-Anderson JL, Worley JF, Dutky SR, Robbins WE, Lusby W (1979) Jour Org Chema *44,* 5002
411. Humphries EC (1968) Field Crop Abstr *21,* 91
412. Wittwer SH (1971) Outlook on Agriculture *6,* 205
413. Humphries EC, Welbank PJ, Witts KJ (1965) Ann Appl Biol *56,* 351
414. Ryerson DK, Oplinger ES, Gritton ET (1978) Jour Amer Soc Hort Sci *103,* 794

415. Behrendt S, Schott PE, Jung J, Bleiholder H, Lang H (1978) Landwirtsch Forsch *35*, 277
416. Schott PE (1979) Med Fac Landbouww Rijksuniv Gent *44*, 853
417. Elkins DM, Vandeventer JW, Briskovich MA (1977) Agron Jour *69*, 458
418. Elkins DM, Suttner DL (1974) Agron Jour *66*, 487
419. Elkins DM, Tweedy JA, Suttner DL (1974) Agron Jour *66*, 492
420. Aagesen GJ, Elkins DM (1976) Agron Jour *68*, 886
421. Elkins DM (1974) Agron Jour *66*, 426
422. Elkins DM, Vandeventer JW, Kapusta G, Anderson MR (1979) Agron Jour *71*, 101
423. Parups EV, Cordukes WE (1977) HortScience *12*, 225
424. Wu CH, Myers HR, Santelmann PW (1976) Agron Jour *68*, 949
425. Watschke TL, Long FW, Duich JM (1979) Weed Sci *27*, 224
426. Watschke TL, Wehner DJ, Duich JM (1977) Proc NE Weed Sci Soc *31*, 378
427. Watschke TL (1976) Agron Jour *68*, 787
428. Jagschitz JA (1976) Proc NE Weed Sci Soc *30*, 327
429. Hield H, Hemstreet S, Gibeault VA, Youngner VB (1979) Calif Agri *33*(10), 15
430. Engel RE, Aldrich RJ (1955) Proc NE Weed Control Conf *9*, 353
431. Goss RL, Zook F (1971) Golf Superintendent *39*, 46
432. Elkins DM (1977) Grounds Maintenance *12*, 70
433. Elkins DM (1972) Weeds Trees, and Turf *11* (5), 18
434. Coorts GD, Elkins DM (1976) Abstr. 1976 Mtg. Plant Growth Regulator Working Group, p 6
435. Graham BA, Puttock MA, Felauer EE, Neidermyer RW US 4 127 402 (10.6.77/28.11.78)
436. Fridinger TL: US 4 013 444 (19.6.75/22.3.77)
437. Chappell WE, Coartney JS, Link ML (1977) So Weed Sci Soc *13*, 300
438. Glenn S, Rieck CE, Ely DG, Bush LP (1980) Jour Agri Food Chem *28*, 391
439. Schott PE, Nolle HH, Will H (1978) Rasen, Grünflächen, Begrünungen *9*, 39
440. Howell SL, Long CE (1976) Abstr 1976 Mtg. Plant Growth Regulator Working Group, p 6
441. Hamm PC: US 3 850 608 (30.10.72/26.11.74)
442. Thomas GJ: US 4 094 664 (4.2.76/13.6.78)
443. de Silva WH, Bocion PF, Eggenberg P, de Mur A (1979) Z Pflanzenkrankheiten und Pflanzenschutz *86*, 546
444. Schneider G (1970) Ann Rev Plant Physiol *21*, 499
445. Maestri N, Currier HG (1958) Weeds *6*, 315
446. Turgeon AJ, Meggitt WF (1971) Proc NE Weed Sci Soc *25*, 399
447. Turgeon AJ, Meggitt WF, Penner D (1972) Weed Sci *20*, 562
448. Plant HL, Zukel JW, Ames RB: US 4 120 692 (20.7.77/17.10.78)
449. Ehrenfreund J: US 4 164 403 (6.6.78/14.8.79)
450. Dicks JW (1976) Outlook on Agriculture *9*, 69
451. Dicks JW (1972) Ann Appl Biol *72*, 313
452. Menhenett R (1977) Ann Appl Biol *87*, 451
453. Hüppi GA, Bocion PF, de Silva WH (1976) Experientia *32*, 37
454. Menhenett R (1976) Grower *85*, 410
455. Wilfret GJ (1978) Proc Fla State Hort Soc *91*, 220
456. Wilfret GJ, Harbaugh BK, Nell TA (1978) HortScience *13*, 701
457. Lumis GD, Johnson AG (1979) Can Jour Plant Sci *59*, 1161
458. Bushong JW, Gates DW, Sullivan TD (1977) Proc. 1976 Brit. Crop Prot. Conf. (Weeds), p 695
459. Jung J, Dressel J (1977) Z Pflanzenernahr Bodenkd *140*, 375
460. Cothren JT, Nester PR, Stutte CA (1977) Proc Plant Growth Regulator Working Group *4*, 204
461. Gausman HW, Rittig FR, Namken LN, Rodriguez RR, Escobar DE, Garza MV (1978) Proc Plant Growth Regulator Working Group *5*, 137
462. Schott PE, Schroeber M (1979) Proc Plant Growth Regulator Working Group *6*, 250
463. Huglin P (1955) Compt Rend Acad Agri (France) *41*, 1709
464. Vega J, Mavrich EP (1959) Dev Invest Agri (Buenos Aires) *13*, 183

465. Klenert M (1974) Vitis *13,* 8
466. Weaver RJ, Pool BM (1971) Amer Jour Enol Viticult *22,* 223
467. Lavee S, Erez A, Shulman Y (1977) Vitis *16,* 89
468. Norton RA (1973) Proc West Wash Hort Assn *63,* 152
469. Sheets WA (1973) Proc West Wash Hort Assn *63,* 150
470. Waister PD, Cormack NR, Sheets WA (1977) Jour Hort Sci *52,* 75
471. Crandall PC, Chamberlain JD, Garth JKL (1980) Jour Amer Soc Hort Sci *105,* 194
472. Cathey HM (1964) Ann Rev Plant Physiol *15,* 271
473. Furuta T, Jones WC, Mock T, Humphrey W, Maire R, Breece J (1972) Calif Agri *26* (3), 10
474. Furuta T, Jones WC, Humphrey W, Breece J (1972) Florists Dev *150,* 23
475. Henley RW, Poole RT (1974) Proc Fla State Hort Soc *87,* 435
476. Vienravee K, Rogers MN (1974) Florists Dev *155,* 24
477. Tso TC, Steffens GL, Engelhaupt ME (1965) Jour Agri Food Chem *13,* 78
478. Wilcox M: US 3 867 452 (30.12.70/18.2.75)
479. Wilcox M: US 3 891 706 (2.10.72/24.6.75)
480. Cooke AR, Starke GR: US 3 880 643 (22.8.72/29.4.75)
481. Wilcox M: US 4 046 809 (16.12.74/6.9.77)
482. Wilcox M, Chen IY, Kennedy PC, Li YY, Kincaid LR, Helseth NT (1977) Proc Plant Growth Regulator Working Group *4,* 194
483. Kupelian RH: US 4 123 250 (4.2.74/31.10.78)
484. Wilcox M: US 4 116 667 (22.6.77/26.9.78)
485. Wilcox M: US 4 169 721 (5.12.77/2.10.79)
486. Wilcox M, Whitty EB, Chen IY, Li YY, Clark F, Kennedy PC, Kincaid LR, Helséth NT, Hensley JR (1979) Proc Plant Growth Regulator Working Group *6,* 96
487. Wilcox M, Whitty EB, Li YY, Chen IY, Kincaid LR, Hensley JR, Kennedy PC (1978) Proc Plant Growth Regulator Working Group *5,* 167
488. Kennedy PC, Seltmann H, Atkinson WO, Whitty EB, Wilcox M (1978) Proc Plant Growth Regulator Working Group *5,* 172
489. Kennedy PC (1979) Proc Plant Growth Regulator Working Group *6,* 102
490. Steffens GL (1980) Tobacco Sci *24,* 102
491. Ogata Y, Fukuhara T: US 4 182 621 (25.5.78/8.1.80)
492. Yu PK: US 4 124 370 (8.5.75/7.11.78)
493. Collins DJ: US 4 155 743 (24.4.78/22.5.79)
494. Ashkar SAK: US 4 090 860 (9.9.77/23.5.78)
495. Ashkar SAK: US 4 093 441 (9.9.77/6.6.78)
496. Wall ME, Wani MC, Cook CE, Palmer KH, McPhail AT, Sim GA (1966) Jour Amer Chem Soc *88,* 3888
497. Worley JF, Spaulding DW, Buta JG (1979) Tobacco Sci *23,* 43
498. Sachs RM, Hackett WP (1972) HortScience *7,* 440
499. Cathey HM, Steffens GL, Stuart NW, Zimmerman RH (1966) Science *153,* 1382
500. Sill LZ, Nelson PV (1970) Jour Amer Soc Hort Sci *95,* 270
501. McDowell T (1967) Ohio Florists' Assn Bull *455,* 1
502. Larson RA, McIntyre M (1967) Florists' Rev *141,* 21
503. Bocion PF, de Silva WH, Hüppi GA, Szkrybalo W (1975) Nature *258,* 142
504. de Silva WH, Bocion HR, Walther HR (1976) HortScience *11,* 569
505. Sanderson KC, Martin WC (1977) HortScience *12,* 337
506. Larson RA (1978) Jour Hort Sci *53,* 57
507. Orson P, Kofranek AM (1978) Jour Amer Soc Hort Sci *103,* 801
508. Sanderson KC, Martin WC (1978) Proc So Nurserymens' Assn Res Conf *23,* 152
509. Breece JR, Furuta T, Hield HZ (1978) Univ. Calif. Crop Ext. Flower and Nursery Rpt. Winter 1978, p 1
510. Shu LJ, Sanderson KC (1979) Proc So Nurserymens' Assn Res Conf *24,* 201
511. Larson RA, Hilliard BG (1979) No Carolina Flower Growers' Bull *23* (2), 4
512. Adriansen E (1979) Tidsskrift for Planteavl *83,* 205
513. Arzee T, Landenauer H, Gressel J (1977) Bot Gaz *138,* 18

514. Bocion PF, de Silva WH, Walther HR, Graf HR (1978) Brit Crop Prot Coun Monogr *21,* 195
515. de Silva WH, Bocion PF, Walther HR, Graf HR (1977) Brit Plant Growth Regulator Group News Bull *1,* 8
516. Johnson AG, Lumis GP (1979) HortScience *14,* 626
517. Sachs RM, Hield HZ, deBie J (1975) HortScience *10,* 367
518. Malstrom HL, McMeans JL (1977) HortScience *12,* 68
519. Martin GC, LaVine P, Sibbett GS, Nishijima C (1980) Calif Agri *34* (10), 16
520. Worley RE (1980) HortScience *15,* 180
521. Tucker DJ, Maw GA (1975) Sci Hort *3,* 331
522. Maw GA (1977) Sci Hort *7,* 43
523. Maw GA, Tucker DJ (1979) Grower *91* (23), 18
524. Vereecke M (1975) HortScience *10,* 420
525. Vereecke M, Benoit F (1976) Med Fac Landbouww Rijksuniv Gent *41,* 1087
526. Wilfret GJ, Harbaugh BK (1976) HortScience *11,* 304
527. Lewis AJ (1980) HortScience *15,* 310
528. Tjia B, Johanson S, Buxton J (1977) HortScience *12,* 259
529. Quinlan JD (1978) Acta Hort *80,* 39
530. Quinlan JD, Preston AP (1973) Acta Hort *34,* 123
531. Quinlan JD (1978) Acta Hort *65,* 129
532. Baldini E, Sansavini S, Zocca A (1973) Jour Hort Sci *48,* 327
533. Villemur P, Suquet J, Marger J (1976) Pomol Franc *18,* 65
534. Quinlan JD, Preston AP (1978) Jour Hort Sci *53,* 39
535. Villemur P, Marger J, Massol R (1979) Fruit Belge *47,* 334
536. Anonymous (1979) The Grower, October *4,* 46
537. England DJF (1978) Brit Crop Prot Coun Monogr *21,* 203
538. Larsen FE (1979) Jour Amer Soc Hort Sci *104,* 770
539. Nickell LG (1977) Ecophysiology of Tropical Crops. In: Alvim P de T, Kozlowski TT (eds), Academic Press, New York
540. Leopold AC (1949) Amer Jour Bot *36,* 437
541. Jewiss OR (1972) Jour Brit Grassland Soc *27,* 65
542. Langer RHM, Prasad DC, Laude HM (1973) Ann Bot *37,* 565
543. Johnston GFS, Jeffcoat B (1977) New Phytol *79,* 239
544. Saini AD, Sud A, Nanda R (1975) Ind Jour Plant Physiol *18,* 140
545. Chang WC (1979) NSC Symp Ser *2,* 123
546. Eastwood D (1979) Trop Agri *56,* 11
547. Takahashi DT (1969) Ann. Rpt. Exp. Stn. Hawaiian Sugar Planters' Assn, p 50
548. Lucchesi AA, Florencio AC, Godoy OP, Stupiello JP (1979) Bras Acucareiro *93,* 19
549. Ingram JW, Bynum EK, Charpentier LJ (1947) Jour Econ Entomol *40,* 745
550. Adams JB (1960) Can Jour Zool *38,* 285
551. Robinson AG, (1959) Can Entomol *91,* 527
552. Robinson AG (1960) Can Entomol *92,* 494
553. Robinson AG (1961) Can Jour Plant Sci *41,* 413
554. Maxwell RC, Harwood RF (1958) Bull Entomol Soc Amer *4,* 100
555. Maxwell RC, Harwood RF (1960) Ann Entomol Soc Amer *53,* 199
556. Van Emden HF (1964) Nature *201,* 946
557. Honeyborne CHB (1969) Jour Sci Food Agri *20,* 388
558. Eichmeier J, Guyer G (1960) Jour Econ Entomol *53,* 661
559. Rodriguez JG, Campbell JM (1960) Jour Econ Entomol *54,* 984
560. Tahori AS, Zeidler G, Halevy AH (1965) Jour Sci Food Agri *16,* 570
561. Carlisle DB, Osborne DJ, Ellis PE, Moorhouse JE (1963) Nature *200,* 1230
562. Ellis PE, Carlisle DB, Osborne DJ (1965) Science *149,* 546
563. Carlisle DB, Ellis PE, Osborne DJ (1969) Jour Sci Food Agri *20,* 391
564. El-Ibrashy MT, Mansour MH (1970) Experientia *26,* 1095
565. Dimetry NZ, Mansour MH (1975) Z Pflanzenkrankh Pflanzenschutz *82,* 561
566. Adkisson PL, Wilkes LH, Johnson SD (1958) Tex Agr Exp Stn Bull *920,* 1
567. Adkisson PL (1961) Jour Econ Entomol *54,* 1107

568. Adkisson PL (1962) Jour Econ Entomol *55,* 949
569. Kittock DL, Mauney JR, Arle HF, Bariola LA (1973) Jour Environ Qual *2,* 405
570. Kittock DL, Arle HF, Bariola LA (1975) Proc. Beltwide Cotton Prod. Res. Conf., p 71
571. Kittock DL, Arle HF, Bariola LA, Vail PV (1975) Proc. Western Cotton Prod. Conf., p 59
572. Bariola LA, Kittock DL, Arle HF, Vail PV, Henneberry TJ (1976) Jour Econ Entomol *69,* 633
573. Kittock DL, Arle HF (1976) Proc. Beltwide Cotton Res. Prod. Conf., p 48
574. Kittock DL, Arle HF (1977) Crop Sci *17,* 320
575. Kittock DL, Arle HF, Henneberry TJ, Bariola LA (1978) USDA, ARS W–52, 1
576. Kittock DL, Henneberry TJ, Bariola LA (1979) Proc. Beltwide Cotton Prod. Res. Conf., p 62
577. Kittock DL, Arle HF, Henneberry TJ, Bariola LA, Walhood VT (1980) Crop Sci *20,* 330
578. Henneberry TJ, Bariola LA, Kittock DL (1980) USDA Tech Bull 1610, 1
579. Thomas RO, Cleveland TC, Cathey GW (1979) Crop Sci *19,* 861
580. Davis D, Dimond AE (1953) Phytapath *53,* 137
581. Sinha AK, Wood RKS (1967) Ann Appl Biol *59,* 117
582. Oswald TH, Wyllie TD (1973) Plant Dis Rptr *57,* 789
583. Daft MJ (1965) Ann Appl Biol *55,* 51
584. Milo GE, Srivastava BS (1969) Virology *38,* 26
585. Fletcher RA, Quick WA, Phillips DR (1968) In: Biochemistry and Physiology of Plant Growth Substances, Runge Press, Ottawa, p 1447
586. Kiraly Z, Szirmai J (1964) Virology *23,* 286
587. Fraser RSS, Whenham RJ (1978) Physiol Plant Path *13,* 51
588. Selman IW (1964) Ann Appl Biol *53,* 67
589. Aldwinckle HS, Selman IW (1967) Ann Appl Biol *60,* 49
590. Aldwinckle HS (1975) Virology *66,* 341
591. Bailiss KW, Cocker FM, Cassells AC (1977) Ann Appl Biol *87,* 383
592. Buchenauer H, Erwin DC (1976) Phytopathol *6,* 1140
593. Erwin DC, Tsai SD, Khan RA (1976) Phytopathol *66,* 106
594. Erwin DC, Tsai SD, Khan RA (1979) Phytopathol *69,* 283
595. Erwin DC, Tsai SD, Khan RA (1979) Calif Agri *33* (4), 8
596. McIntosh AH, Bateman GL (1979) Ann Appl Biol *92,* 29
597. Zevite-Kulvetiene Z, Baneviciene Z (1979) Zashch Rast (Moscow) *5,* 32
598. Rawlins TE (1962) Plant Dis Rptr *46,* 170
599. Halevy AH, Kessler B (1963) Nature *197,* 310
600. Marth PC, Ray JR (1961) Jour Agri Food Chem *9,* 359
601. Plaut Z, Halevy AH, Shmueli E (1964) Israel Jour Agri Res *14,* 153
602. Dolgopolova LN, Lakhanov AP (1979) Khim Sel'sk Khoz *17* (9), 27
603. Livne A, Vaadia Y (1965) Physiol Plant *18,* 658
604. Meidner H (1967) Jour Exp Bot *18,* 556
605. Luke HH, Freeman TG (1968) Nature *217,* 873
606. Mizrahi Y, Richmond AE (1972) Aust Jour Biol Sci *25,* 437
607. Vaadia Y (1976) Phil Trans Royal Soc London *B273,* 513
608. Little CHA, Eidt PC (1968) Nature *220,* 498
609. Mittelheuser CJ, Van Steveninck RFM (1969) Nature *221,* 281
610. Jones RJ, Mansfield TA (1970) Jour Exp Bot *21,* 714
611. Mizrahi Y, Blumenfeld A, Richmond AE (1970) Plant Physiol *46,* 169
612. Hiron RWP, Wright STC (1973) Jour Exp Bot *24,* 769
613. Davies WJ, Kozlowski TT (1975) Forest Sci *21,* 191
614. Davies WJ, Kozlowski TT (1975) Can Jour Forest Res *5,* 90
615. Raschke K (1976) Phil Trans Royal Soc London *B273,* 551
616. Zelitch I, Waggoner PE (1962) Proc Nat'l Acad Sci (USA) *48,* 1101
617. Zelitch I, Waggoner PE (1962) Proc Nat'l Acad Sci (USA) *48,* 1297
618. Shimshi D (1963) Plant Physiol *38,* 709
619. Shimshi D (1963) Plant Physiol *38,* 713
620. Slatyer RO, Bierhuizen JF (1964) Aust Jour Biol Sci *17,* 131

621. Davenport DC (1966) Nature *212,* 801
622. Davenport DC (1967) Jour Exp Bot *18,* 332
623. Waisel Y, Borger GA, Kozlowski TT (1969) Plant Physiol *44,* 685
624. Das VSR, Raghavendra AS (1979) Outlook on Agriculture *10,* 92
625. Mizrahi Y, Scherings HG, Malis Arad S, Richmond AE (1974) Physiol Plant *31,* 44
626. Erkan Z, Bangerth F (1980) Angew Bot *54,* 207
627. Raschke K (1974) Plant Physiol *53,* (Suppl.), 55
628. Santakumari M, Reddy CS, Das VSR (1977) Proc Ind Acad Sci *B86,* 143
629. Zelitch I (1964) Science *143,* 692
630. Orchard PW (1977) Jour Australian Inst Agri Sci *43,* 68
631. Mishra D, Pradhan GC (1972) Plant Physiol *50,* 271
632. Raghavendra AS, Das VSR (1977) Jour Exp Bot *28,* 480
633. Stoddard EM, Miller PM (1962) Science *137,* 224
634. Das VSR, Rao IM, Swamy PM (1977) Ind Jour Exp Biol *15,* 642
635. Davies WJ, Kozlowski TT (1974) Jour Amer Soc Hort Sci *99,* 297
636. Gale J, Hagan RM (1966) Ann Rev Plant Physiol *17,* 269
637. Zelitch I (1969) Ann Rev Plant Physiol *20,* 329
638. Rojas-Garciduenas M, Gamez H (1978) Turrialba *28,* 307
639. Caceres JR, Rojas-Garciduenas M (1980) Turrialba *30,* 25
640. Cathey HM (1964) Ann Rev Plant Physiol *15,* 271
641. Marth PC (1965) Jour Agri Food Chem *13,* 331
642. Wunsche V (1966) Naturwissenschaften *53,* 386
643. Birecka H, Zebrowski Z (1966) Bull Acad Pol Sci *14,* 367
644. Irving MR, Lanphear FO (1968) Plant Physiol *43,* 9
645. Granger RL, Hogue EJ (1968) Can Jour Plant Sci *48,* 100
646. Rikin A, Richmond AE (1976) Physiol Plant *38,* 95
647. Rikin A, Blumenfeld A, Richmond AE (1976) Bot Gaz *137,* 307
648. Rikin A, Atsmon D, Gitler C (1979) Plant Cell Physiol *20,* 1537
649. Villareal RL (1980) Linking Research to Crop Production. In: Staples, RC, Kuhr RJ (eds) Plenum New York
650. Okii M, Onitake T, Kawai M, Takematsu T, Konnai M: US 4 231 789 (5.10.78/4.11.80)
651. Riehl LA, Coggins CW, Carman GE (1966) Jour Econ Entomol *59,* 615
652. Devlin RM, Karczmarczyk SJ (1977) Proc NE Weed Sci Soc *31,* 156
653. Gruzdev LG (1979) Fiziol Rast *26,* 153
654. Shukla SN, Tewari MN (1974) Ind Jour Agri Sci *44,* 787
655. Knavel DE (1969) Jour Amer Soc Hort Sci *94,* 32
656. Castro PRC, Malavolta E (1977) Turrialba *27,* 273
657. Maier VP, Yokoyama H (1974) Nutritional Qualities of Fruits and Vegetables. In: White PL, (ed) Futura, New York
658. Yokoyama H, Hsu WJ, Poling SM, Hayman E, DeBenedict C (1977) Proc Internat'l Soc Citriculture *3,* 717
659. Coggins CW, Henning GL, Yokoyama H (1970) Science *168,* 1589
660. Knypl JS (1969) Naturwissenschaften *56,* 572
661. Yokoyama H US 3 684 530 (25.5.71/15.8.72)
662. Poling SM, Hsu WJ, Yokoyama H (1973) Phytochem *12,* 2665
663. Poling SM, Hsu WJ, Yokoyama H (1975) Phytochem *14,* 1933
664. Poling SM, Hsu WJ, Yokoyama H (1976) Phytochem *15,* 1685
665. Coggins CW, Jones WW (1977) Proc Internat'l Soc Citriculture *2,* 686
666. Coggins CW, Hall AE (1975) Jour Amer Soc Hort Sci *100,* 484
667. Hall AE, Coggins CW (1978) Physiol Plant *44,* 221
668. Buescher RW (1977) HortScience *12,* 315
669. Buescher RW, Doherty JH (1978) Jour Food Science *43,* 1816
670. Worku W, Harner RC (1971) HortScience *6,* 279
671. Knavel DE, Kemp TR (1973) HortScience *9,* 403
672. Cantliffe DJ, Goodwin P (1975) Jour Amer Soc Hort Sci *100,* 157
673. Locascio SJ, Smith TS (1977) Proc Fla State Hort Soc *90,* 421
674. Nakayama RM, Matta FB (1973) HortScience *8,* 252

675. Lockwood D, Vines HM (1972) Jour Amer Soc Hort Sci *97,* 192
676. Bramlage WJ, Devlin RM, Smagula JM (1972) Jour Amer Soc Hort Sci *97,* 625
677. Eck P (1972) Jour Amer Soc Hort Sci *97,* 213
678. Rigbey B, Dana MN, Binning LK (1972) HortScience *7,* 82
679. Shawa AY (1979) HortScience *14,* 168
680. Jensen F, Andris H (1977) Calif Agri *31* (8), 18
681. Powers JR, Shively EA, Nagel CW (1980) Amer Jour Enol Viticulture *31,* 203
682. Gorini FL, Mariotti A (1973) Informatore Agrario *29,* 13871
683. Modlibowska I, Wickenden MF (1974) 1973 Rpt. East Malling Res. Stn, p 59
684. Strydom DK, Honeyborne GE (1975) Deciduous Fruit Grower *25,* 112
685. Johnson DS, Sharples RO, Cappelini P (1975) 1974 Rpt. East Malling Res. Stn., p 77
686. Greene DW, Lord WJ (1975) HortScience *10,* 254
687. Child RD, Atkins H (1975) 1974 Rpt. Long Ashton Res. Stn., p 28
688. Hammett LK (1976) HortScience *11,* 57
689. Jones KM (1979) Aust Jour Exp Agri Animal Husbandry *19,* 251
690. Batjer LP, Williams MW, Martin GC (1964) Proc Amer Soc Hort Sci *85,* 11
691. Ryugo K (1966) Proc Amer Soc Hort Sci *88,* 160
692. Looney NE (1969) Can Jour Plant Sci *49,* 625
693. Proebsting EL, Mills HH (1976) Jour Amer Soc Hort Sci *101,* 175
694. Drake SR, Proebsting EL, Thompson JB, Nelson JW (1980) Jour Amer Soc Hort Sci *105,* 668
695. Moshonas MG, Shaw PE (1978) Jour Agri Food Chem *26,* 1288
696. Domir SC, Foy CL (1976) Tobacco Sci *20,* 158
697. Steffens GL, Alphin JG, Ford ZT (1970) Beitr Tabakforsch *5,* 262
698. Nickell LG, Tanimoto TT (1965) Rpts. 25th Ann. Conf. Haw. Sugar Technol., p 152
699. Hurney AP, Schmalzl K (1978) Proc. 45th Conf. Queensland Soc. Sugar Cane Technol., p 139
700. Kingston G, Chapman LS, Hurney AP (1978) Proc. 45th Conf. Queensland Soc. Sugar Cane Technol., p 37
701. Chapman LS, Kingston G (1977) Proc Queensland Soc Sugar Cane Technol *44,* 143
702. Bieske GC (1970) Proc Queensland Soc Sugar Cane Technol *37,* 117
703. Vlitos AJ, Lawrie ID (1967) Proc. 12th Cong. Internat'l. Soc. Sugar Cane Technol., p 429
704. Vlitos AJ, Fewkes DW (1969) Sugar y Azucar *64* (9), 27
705. Humbert RP (1974) World Farming *16* (12), 25
706. Julien HR, McIntyre G (1970) Ann Rpt Mauritius Sugar Ind Res Inst *18,* 130
707. Julien MHR (1974) Exp Agri *10,* 113
708. Julien MHR (1974) Exp Agri *10,* 123
709. Julien HR, Goolambossen M (1976) Rev Agri Suc Ile Maurice *55,* 389
710. Julien HR, Soopramanien GC, Martine JF, Medan H (1978) Rev Agri Suc Ile Maurice *57,* 172
711. Anonymous (1972) Ann. Rpt. 1971–72 So. Afr. Sugar Assn. Exp. Stn., p 32
712. Rostron H (1977) Proc. 51st Cong. So. Afr. Sugar Technol. Assn., p 30
713. Rostron H (1977) Proc. 16th Cong. Internat'l. Soc. Sugar Cane Technol., vol 2, p 1605
714. Sweet CPM (1977) Proc. 16th Cong. Internat'l. Soc. Sugar Cane Technol., vol 2, p 1619
715. Clowes MSJ, Wood RA (1978) Proc. 52nd Ann. Cong. So. Afr. Sugar Technol. Assn., p 166
716. Srivastava SC, Singh B, Singh K (1971) Proc. Sugar Technol. Assn. India *4,* A–1
717. Chacravarti AS, Sarkar AK, Thakur AK (1975) Proc. 5th Joint Conv. Indian Sugar Technol. Assns., p 67
718. Kumar A, Narasimhan R (1977) Indian Sugar *26,* 817
719. Rao KC, Asokan S (1977) Proc. 6th Joint Conv. Indian Sugar Technol. Assns., p 133
720. Chacravarti AS, Thakur AK, Sarkar AK (1977) Proc. 6th Joint Conv. Indian Sugar Technol. Assns., p 167
721. Sharma RA, Sharma RK, Sharma SR (1977) Indian Sugar Crops Jour *4,* 35
722. Azzi GM, Alves AS, Kumar A (1977) Proc. 16th Cong. Internat'l. Soc. Sugar Cane Technol., vol 2, p 1653

723. Alves AS, Azzi GM, Kumar A (1977) Proc. 16th Cong. Internat'l. Soc. Sugar Cane Technol., vol 2, p 1713
724. Tianco AP, Escober TR (1970) Proc. 18th Conv. Philsutech, p 93
725. Zamora OB, Rosario EL (1977) Phil Jour Crop Sci *2,* 133
726. Gonzales MY, Tianco AP (1978) Victorias Agri Res Rpts *16,* 18
727. Yang PC, Ho FW (1976–77) Ann. Rpt. Taiwan Sugar Res. Inst., p 8
728. Yang PC, Ho FW (1977) Proc. 16th Cong. Internat'l. Soc. Sugar Cane Technol., vol 2, p 1701
729. Yang PC, Ho FW (1978) Taiwan Sugar *25* (3), 101
730. Yang PC, Ho FW (1979) Taiwan Sugar *26* (1), 30
731. Yates RA, Bates JF (1957) Proc. 1957 Mtg. Brit. West Indies Sugar Technol., p 174
732. Evans H, Bates JF (1962) Proc. 11th Cong. Internat'l. Soc. Sugar Cane Technol., p 298
733. Jaramillo H, Schuitemaker F, Garcia G (1977) Proc. 16th Cong. Internat'l. Soc. Sugar Cane Technol., vol 2, p 1931
734. Samuels G, Velez A, Yates RA, Walker B (1972) Jour Agri Univ Puerto Rico *56,* 370
735. Alexander AG, Montalvo-Zapata R (1973) Trop Agri (Trinidad) *50,* 35
736. Alexander AG, Montalvo-Zapata R (1973) Trop Agri (Trinidad) *50,* 307
737. Holder DG, De Stefano RP (1979) Sugar Jour *41* (9), 21
738. Andreis HJ, De Stefano RP (1979) Sugar Jour *41* (11), 21
739. Andreis HJ, De Stefano RP (1980) Sugar Jour *43* (1), 26
740. Hamm PC: US 3 535 603 (21.10.68/25.8.70)
741. Thomas GJ: US 4 094 664 (4.2.76/13.6.78)
742. Otten GG: US 4 120 688 (15.8.77/17.10.78)
743. Nickell LG: US 3 992 187 (25.2.75/16.11.76)
744. Nickell LG (1977) Amer Chem Soc Adv Chem Series No *159,* 6
745. Quebedeaux B: US 3 619 166 (1.4.69/9.11.71)
746. Nickell LG: US 4 033 755 (30.3.76/5.7.77)
747. Leach RWA: US 4 056 385 (16.11.76/1.11.77)
748. Nickell LG, Takahashi DT (1971) Rpts. 30th Ann. Conf. Haw. Sugar Technol., p 73
749. Yates RA (1972) Trop Agri (Trinidad) *49,* 235
750. Nickell LG, Takahashi DT (1972) Rpts. 31st Ann. Conf. Haw. Sugar Technol., p 47
751. Nickell LG, Takahashi DT (1973) Rpts. 32nd Ann. Conf. Haw. Sugar Technol., p 76
752. Nickell LG (1974) Bull Plant Growth Regulators *2,* 51
753. Osgood RV (1977) Rpts. 36th Ann. Conf. Haw. Sugar Technol., p 60
754. Pan YC, Lee YP (1977) Proc. 16th Cong. Internat'l. Soc. Sugar Cane Technol., vol 2, p 1693
755. Zschoche WC (1977) Sugar y Azucar *72* (4), 21
756. Osgood RV, Teshima A (1979) Proc. Plant Growth Reg. Work Group *6,* 29
757. Rueppel ML: US 4 047 926 (3.5.76/13.9.77)
758. Evans AW: US 3 291 592 (20.4.64/13.12.66)
759. Porter CA: US 3 909 233 (4.3.74/30.9.75)
760. Nickell LG, Tanimoto TT (1968) Rpts. 28th Ann. Conf. Haw. Sugar Cane Technol., p 104
761. Nickell LG, Tanimoto TT: US 3 493 361 (11.1.68/3.2.70)
762. Nickell LG, Maretzki A (1970) Haw Planters' Rec *58* (5), 71
763. Vega N (1971) Bol Est Exp Occidente (Venezuela) *93,* 3
764. Carlson AE: US 3 224 865 (20.11.62/21.12.65)
765. Gaertner VR, Hamm PC: US 4 063 922 (13.8.76/20.12.77)
766. Nickell LG: US 3 671 219 (20.7.70/20.6.72)
767. Bosshard R, Muller JC: US 3 898 071 (17.1.74/5.8.75)
768. Nickell LG: US 3 992 190 (17.4.75/16.11.76)
769. Jaworski EG: US 3 981 718 (20.12.74/21.9.76)
770. Nickell LG: US 3 660 072 (22.5.70/2.5.72)
771. Porter CA: US 3 826 641 (12.2.73/30.7.74)
772. Nickell LG: US 3 994 715 (17.4.75/30.11.76)
773. Weakley ML: US 3 860 411 (27.7.72/14.1.75)
774. Nickell LG, Tanimoto TT: US 3 505 056 (18.12.67/7.4.70)

775. Nickell LG, Takahashi DT (1975) Haw Planters' Rec *59* (2), 15
776. Nickell LG, Tanimoto TT: US 3 482 958 (5.4.67/9.12.69)
777. Vega N (1971) Bol Est Exp Occidente (Venezuela) *93,* 33
778. George EF, Phillips MR (1978) Brit Crop Prot Coun Monogr No *21,* 211
779. Nickell LG: US 3 704 111 (14.8.70/28.11.72)
780. Nickell LG: US 4 099 957 (25.5.77/11.7.78)
781. Levitt G: US 4 190 432 (1.8.77/26.2.80)
782. Nickell LG, Tanimoto TT: US 3 482 959 (6.3.67/9.12.69)
783. Nickell LG: US 3 870 503 (3.10.73/11.3.75)
784. Nickell LG: US 3 897 239 (28.3.74/29.7.75)
785. Garrod JF, Wells WH (1978) Brit Crop Prot Coun Monogr No *21,* 217
786. Copping LG, Garrod JF: US 4 139 365 (31.3.77/13.2.79)
787. Franz JE: US 4 110 100 (12.8.77/29.8.78)
788. Franz JE: US 3 996 040 (27.6.75/7.12.76)
789. Kupelian RH: US 3 874 872 (2.3.73/1.4.78)
790. Osgood RV (1980) Proc Plant Growth Reg Work Group *7,* 149
791. Pulido ML (1974) Sugar y Azucar *69* (6), 105
792. Buckman SJ, Pulido ML: US 3 854 928 (14.11.73/17.12.74)
793. Nickell LG: US 3 992 186 (5.4.74/16.11.76)
794. Ratts KW: US 3 961 934 (19.12.74/8.6.76)
795. Pfeiffer RK: US 3 245 775 (22.1.63/12.4.66)
796. Nickell LG: US 3 897 240 (28.3.74/29.7.75)
797. Guyot HM: US 3 307 932 (12.5.64/7.3.67)
798. Nickell LG: US 3 930 840 (5.4.74/6.1.76)
799. Nickell LG: US 3 909 238 (15.4.74/30.9.75)
800. Ahlrichs LE, Porter CA (1972) Proc. 11th Brit. Weed Control Conf., p 1215
801. Rice ER, Holder DG, De Stefano RP (1980) Sugar Jour *43* (5), 23
802. Chapman GW (1951) Jour Rubber Res Inst Malaya *13,* 167
803. Baptiste EDC, de Jonge P (1955) Jour Rubber Res Inst Malaya *14,* 362
804. Abraham PD, Tayler RS (1967) Exp Agri *3,* 1
805. Tupy J (1969) Planta *88,* 144
806. Abraham PD, Blencowe JW, Chua SE, Gomez JB, Moir GFJ, Pakianathan SW, Sekhar BC, Southorn WA, Wycherley PR (1971) Jour Rubber Res Inst Malaya *23* (2), 85
807. Dickenson PB (1976) Outlook on Agriculture *9,* 88
808. Abraham PD, Blencowe JW, Chua SE, Gomez JB, Moir GFJ, Pakianathan SW, Sekhar BC, Southorn WA, Wycherley PR (1971) Jour Rubber Res Inst Malaya *23* (2), 114
809. Abraham PD (1970) Plant Bull Rubber Res Inst Malaya *111,* 366
810. Abraham PD, Blencowe JW, Chua SE, Gomez JB, Moir GFJ, Pakianathan SW, Sekhar BC, Southorn WA, Wycherley PR (1971) Jour Rubber Res Inst Malaya *23* (2), 90
811. Coupe M, d'Auzac J (1974) Physiol Veg *12,* 1
812. Tupy J, Primot L (1976) Biol Plantarum (Praha) *18,* 373
813. Low FC (1978) Jour Rubber Res Inst Malaysia *26* (1), 21
814. Chandrasekera LB (1973) Jour Rubber Res Inst Sri Lanka *50,* 12
815. Mainstone BJ, Ng WC, Wai ST (1977) Jour Rubber Res Inst Sri Lanka *54,* 131
816. Chandrasekera LB (1977) Jour Rubber Res Inst Sri Lanka *54,* 125
817. Satchuthananthavale R, Weerasinghe TC (1977) Jour Rubber Res Inst Sri Lanka *54,* 150
818. Moraes VHF (1978) Pesquisa Agropecuaria Brasil *13* (4), 17
819. Moraes VHF, Neto OG, Seeschaaf KW (1978) Pesquisa Agropecuaria Brasil *13* (4), 27
820. George MJ, Pillai VB, Nair VKB, Sethuraj MR (1975) Rubber Board India Bull *12* (3), 98
821. Bridge K (1980) Proc Plant Growth Reg Work Group *7,* 207
822. Pakianathan SW, Jaffar H, Ghani A (1978) Plant Bull Rubber Res Inst Malaysia *155,* 61
823. Chacko EK, Randhawa GS, Menon MA, Negi SP (1972) Curr Sci (India) *41,* 455
824. Shanmugavelu KG, Chittiraichelvan R, Madhava Rao VN (1976) Jour Hort Sci *51,* 425
825. Yokoyama H, Hayman EP, Hsu WJ, Poling SM, Bauman AJ (1977) Science *197,* 1076
826. Yokoyama H, Hayman EP, Hsu WJ, Poling SM: US 4 204 859 (31.3.78/27.5.80)

827. Anonymous (1978) Agri Res *27* (2), 8
828. Anonymous (1975) Agri Res *23* (7), 6
829. Roberts DR (1973) USDA For. Serv. Res. Note SE–191, p 1
830. Wolter KE (1975) Plant Physiol *56,* 149
831. Brown CL, Nix LE (1975) Forest Sci *21,* 359
832. Schwarz OJ, Ryan FJ (1980) Proc Plant Growth Reg Work Group *7,* 199
833. Kidd F, Reid CPP (1979) Forest Sci *25,* 569
834. Parham MR (1976) Outlook on Agriculture *9,* 76
835. Parham MR (1978) Proc 1978 Brit. Crop. Prot. Conf. Weeds. 785
836. Roberts DR, Peters WJ, Harrington TA, Broomfield J, Crews ER: US 3 839 823 (7.9.73/8.10.74)
837. Brown CL, Enos HI: US 3 971 159 (5.2.75/27.7.76)
838. Drew J: US 3 991 515 (22.3.76/16.11.76)
839. Brown CL: US 4 201 566 (4.8.78/6.5.80)
840. Drew J (1978) Chemistry *51* (2), 17
841. Looney NE (1968) Proc Wash State Hort Assn *64,* 1–R
842. Faust M (1973) Acta Hort *34,* 407
843. Williams MW, Batjer LP, Martin GC (1964) Proc Amer Soc Hort Sci *85,* 17
844. Batjer LP, Williams MW, Martin GC (1966) Proc Amer Soc Hort Sci *88,* 76
845. Bartram RD, Olsen KL, Williams MW (1971) HortScience *6,* 240
846. Edgerton LJ (1968) New York Food Life Sci *1,* 19
847. Looney NE (1971) Jour Amer Soc Hort Sci *96,* 350
848. Looney NE (1973) Acta Hort *34,* 397
849. Forsyth FR, Embree CG, Crowe AD (1975) Can Jour Plant Sci *55,* 107
850. Grauslund J (1976) Tidsskr Planteavl *80,* 893
851. Micke WC, Tyler RH, Yaeger JF (1977) Calif Agri *31* (3), 15
852. Greene DW, Lord WJ, Bramlage WJ (1977) Jour Amer Soc Hort Sci *102,* 491
853. Williams MW, Stahly EA (1969) Jour Amer Soc Hort Sci *94,* 17
854. Stembridge GE, Morrell G (1972) Jour Amer Soc Hort Sci *97,* 464
855. Unrath CR (1974) Jour Amer Soc Hort Sci *99,* 381
856. Ferree DC, Stang EJ, Funt RC (1980) Ohio Agri Res Dev Cntr Res Circ *259,* 7
857. Crane JC, Brooks BM (1952) Proc Amer Soc Hort Sci *59,* 218
858. Crane JC (1956) Proc Amer Soc Hort Sci *67,* 153
859. Byers RE, Emerson FH (1969) Jour Amer Soc Hort Sci *94,* 641
860. Byers RE, Emerson FH, Dostal HC (1972) Jour Amer Soc Hort Sci *97,* 420
861. Baumgardner RA, Stembridge GE, Van Blaricom LO, Gambrell CE (1972) Jour Amer Soc Hort Sci *97,* 485
862. Looney NE (1972) Can Jour Plant Sci *52,* 73
863. Guelfat-Reich S, Ben-Arie R (1975) Jour Amer Soc Hort Sci *100,* 517
864. Mitchell JW, Marth PC (1944) Bot Gaz *106,* 199
865. Frieburg SR (1955) Bot Gaz *117,* 113
866. Dedolph RR, Goto S (1960) Hawaii Farm Sci *8* (3), 3
867. Murashige T, Hamilton RA (1962) Hawaii Farm Sci *11* (2), 3
868. Gane R (1936) Great Brit Dept Sci Indus Res Food Invest Bd Dept 1935, 123
869. Burg SP, Burg EA (1965) Science *148,* 1190
870. Burg SP, Burg EA (1965) Bot Gaz *126,* 200
871. Liu FW (1976) Jour Amer Soc Hort Sci *101,* 222
872. Russo L, Dostal HC, Leopold AC (1968) BioScience *18,* 109
873. Srinivasan C (1971) Madras Agri Jour *58,* 778
874. Khan A, Singh UR, Singh G (1977) Punjab Hort Jour *17,* 84
875. Desai BB, Deshpande PB (1978) Physiol Plant *44,* 238
876. Granger RL, Hogue EJ (1968) Can Jour Plant Sci *48,* 100
877. Crandall PC, Chamberlain JD (1972) Jour Amer Soc Hort Sci *97,* 418
878. Craig DL, Aalders LE (1973) HortScience *8,* 313
879. Jolliffe PA (1975) Can Jour Plant Sci *55,* 429
880. Morris JR, Cawthon DL, Nelson GS, Cooper PE (1978) Jour Amer Soc Hort Sci *103,* 804

881. Lipe JA (1980) HortScience *15,* 585
882. Forsyth FR, Craig DL, Stark R (1977) Can Jour Plant Sci *57,* 1099
883. Howell GS, Stergios BG, Stackhouse SS, Bittenbender HC, Burton CL (1976) Jour Amer Soc Hort Sci *101,* 111
884. Eck P (1970) HortScience *5,* 23
885. Dekazos ED (1976) Proc Fla State Hort Soc *89,* 266
886. Austin ME (1979) Ga Agri Res *20* (4), 8
887. Batjer LP, Williams MW, Martin GC (1964) Proc Amer Soc Hort Sci *85,* 11
888. Looney NE (1969) Can Jour Plant Sci *49,* 625
889. Gil G (1974) Ciencia e Invest Agraria *1,* 187
890. Gil G (1975) Ciencia e Invest Agraria *2,* 77
891. Eaks IL, Jones WW (1959) Calif Citrograph *44* (12), 390
892. Coggins CW, Eaks IL, Hield HZ, Jones WW (1963) Proc Amer Soc Hort Sci *82,* 154
893. Coggins CW, Eaks IL (1964) Calif Citrograph *50* (2), 47
894. Coggins CW, Hield HZ (1962) Proc Amer Soc Hort Sci *81,* 227
895. Coggins CW, Lewis LN (1967) Proc Amer Soc Hort Sci *86,* 272
896. Monselise SP (1976) Israel Jour Bot *25,* 99
897. Monselise SP, Weiser M, Shafir N, Goren R, Goldschmidt EE (1976) Jour Hort Sci *51,* 341
898. Erner Y, Goren R, Monselise SP (1976) Jour Hort Sci *51,* 367
899. Erner Y, Goren R, Monselise SP (1976) Jour Amer Soc Hort Sci *101,* 513
900. Browning GB, Cannell MGR (1970) Jour Hort Sci *45,* 223
901. Rodriguez SJ, Molero JJ (1970) Jour Agri Univ Puerto Rico *54,* 689
902. Nakasone L, Shigeura G, Bullock R (1976) Hawaii Agri Exp Stn Achievement Rpt 1975–76, 5
903. Clowes MStJ (1977) Rhod Jour Agri Res *15,* 79
904. Bramblett J (1977) Progressive Farmer *92,* 18
905. Reeder N (1977) Farm Jour *101,* 20
906. Anonymous (1977) Agrichemical Age *20,* 11
907. Anonymous (1978) Agrichemical Age *21,* 14
908. Anonymous (1979) Agrichemical Age *22,* 10
909. Hatley OE (1970) Ph D Thesis, Purdue University
910. Hatley OE, Hermann L, Collins K, Ohlrogge AJ (1974) Down to Earth *30* (1), 4
911. Hatley OE, Hermann L, Ohlrogge AJ (1977) Proc Plant Growth RegWork Group *4,* 156
912. Oplinger ES, Basabe PJ (1977) Proc Plant Growth Reg Work Group *4,* 145
913. Johnson RR, McKibben GE, Jeppson RG, Jurgens SK (1978) Agron Jour *70,* 626
914. Oplinger ES (1978) Proc Plant Growth Reg Work Group *5,* 100
915. Hicks DR, Lueschen WE, Ford JH, Nelson WW, Overdahl CJ, Evans SD, Peterson RH (1979) Agron Jour *71,* 697
916. Ohlrogge AJ, Roth CB, Fulk-Bringmann SS, Brown HM (1979) Proc Plant Growth Reg Work Group *6,* 14
917. Ohlrogge AJ, Oplinger ES, Abdel-Rahman M, Roth JA, Fulk-Bringmann SS (1980) Proc Plant Growth Reg Work Group *7,* 155
918. Oplinger ES (1980) Proc Plant Growth Reg Work Group *7,* 194
919. Smrz J, Pitrik P (1979) Agrochemia *19,* 21
920. Lukasik S (1970) Roczniki Nauk Rolniczych *96,* 53
921. Bosland JM, Hughes DL, Yamaguchi M (1979) HortScience *14,* 729
922. Weaver RJ (1957) The Blue Anchor *34,* 10
923. Stewart WS, Halsey D, Ching TT (1958) Proc Amer Soc Hort Sci *72,* 165
924. Weaver RJ, McCune SB (1960) Bot Gaz *121,* 155
925. Christodoulou A, Weaver RJ, Pool RM (1968) Proc Amer Soc Hort Sci *92,* 301
926. Hale CR, Coombe BG, Hawker JS (1970) Plant Physiol *45,* 620
927. Weaver RJ, Pool RM (1971) Jour Amer Soc Hort Sci *96,* 725
928. Jensen FL, Kissler JJ, Peacock WL, Leavitt GM (1974) The Blue Anchor *51,* 21
929. Jensen FL, Kissler JJ, Peacock WL, Leavitt GM (1975) Amer Jour Enol Vitic *26,* 79
930. Peacock WL, Jensen F, Else JA, Leavitt G (1977) Amer Jour Enol Vitic *28,* 228
931. Chakrawar VR, Rane DA (1977) Vitis *16,* 97

932. Mehta PK, Chundawat BS (1979) Vitis *18,* 117
933. Kishi M, Tasaki M (1958) Japan Gibberellin Res Assn *2,* 13
934. Clore WJ (1965) Proc Amer Soc Hort Sci *87,* 259
935. Siemer SR, Gordon RS, Nickell LG: US 4 231 788 (27.4.79/4.11.80)
936. Torabi M (1980) HortScience *15,* 521
937. Marth PC, Havis L, Prince VE (1950) Proc Amer Soc Hort Sci *55,* 152
938. Weinberger HH (1951) Proc Amer Soc Hort Sci *57,* 115
939. Gambrell CE, Rhodes WH, Sims ET (1967) So Carolina Agri Exp Stn Res Series *101,* 1
940. Byers RE, Emerson FH (1969) Jour Amer Soc Hort Sci *94,* 641
941. Rom RC, Scott KR (1971) HortScience *6,* 134
942. Gambrell CE, Sims ET, Stembridge GE, Rhodes WH (1972) Jour Amer Soc Hort Sci *97,* 265
943. Byers RE, Emerson FH, Dostal HC (1972) Jour Amer Soc Hort Sci *97,* 420
944. Byers RE, Emerson FH (1973) HortScience *8,* 48
945. Edgerton LJ, Greenhalgh WJ (1969) Jour Amer Soc Hort Sci *94,* 11
946. Byers RE, Dostal HC, Emerson FH (1969) BioScience *19,* 903
947. Stembridge GE, Gambrell CE (1971) Jour Amer Soc Hort Sci *96,* 7
948. Stembridge GE, Raff JW (1973) HortScience *8,* 500
949. Byers RE, Emerson FH, Dostal HC (1972) HortScience *7,* 386
950. Buchanan DW, Hall CB, Biggs RH, Knapp FW (1969) HortScience *4,* 302
951. Knapp FW, Hall CB, Buchanan DW, Biggs RH (1970) Phytochem *9,* 1453
952. Das Gupta DK (1975) Exp Agri *11,* 209
953. Brown RH, Ethredge WJ, King JW (1973) Crop Sci *13,* 507
954. Brown RH, Ethredge WJ (1974) Peanut Sci *1,* 20
955. Bauman RW, Norden AJ (1971) Amer Peanut Res Educ Assn Jour *3,* 75
956. Wynne JC, Baker WR, Rice PW (1974) Agron Jour *66,* 192
957. Daughtry CS, Brown RH, Ethredge WJ (1975) Peanut Sci *2,* 83
958. Wu CH, Santelmann PW (1977) Agron Jour *69,* 521
959. Gorbet DW, Rhoads FM (1975) Agron Jour *67,* 373
960. Hallock DL, Alexander MW (1970) Amer Peanut Res Educ Assn Jour *2,* 22
961. Hartzook A, Goldin E (1970) Israel Jour Agri Res *20,* 169
962. Ketring DL (1977) Agron Jour *69,* 110
963. Krishnamoorthy HN, Khun MJ (1972) Ind Jour Plant Physiol *15,* 21
964. Dilley DR (1969) HortScience *4,* 111
965. Hansen E (1943) Proc Amer Soc Hort Sci *43,* 69
966. Hansen E (1946) Plant Physiol *21,* 588
967. Looney NE (1972) Jour Amer Soc Hort Sci *97,* 81
968. Wang CY, Mellenthin WM (1971) Jour Amer Soc Hort Sci *96,* 122
969. Frenkel C, Klein I, Dilley DR (1968) Plant Physiol *43,* 1146
970. Frenkel C, Dyck R (1973) Plant Physiol *51,* 6
971. Frenkel C, Haard NF (1973) Plant Physiol *52,* 380
972. Wang CY, Mellenthin WM (1977) Plant Physiol *59,* 546
973. Hess PJ, Romani RJ (1980) Plant Physiol *65,* 372
974. Lockwood D, Vines HM (1970) Proc So Agri Workers, Inc *67,* 163
975. Worku Z, Herner RC, Carolus RL (1975) Sci Hort *3,* 239
976. Barreiro M, Kramarovsky E (1975) Fitotec Latinoam *11,* 35
977. Haynes RL, Pothuluri JV (1978) Arkansas Farm Res *27* (6), 15
978. Fogle HW, Blodgett EC, Carter GH, Ingalsbe DW, Neubert AM (1955) Proc Wash State Hort Assn *51,* 109
979. Verner L, Kochan WJ, Loney CE, Moore DC, Kamal AL (1962) Idaho Agri Exp Stn Res Bull *56,* 1
980. Proebsting EL, Mills HH (1966) Proc Amer Soc Hort Sci *89,* 135
981. Leavitt GM, Gerdts MH, Obenauf GL, Mitchell FG, Andris H (1977) Calif Agri *31* (6), 18
982. Dirou JF, Logan BJ (1977) Agri Gaz NSW *88* (6), 34
983. Dirou JF, Logan BJ (1978) Agri Gaz NSW *89* (1), 47
984. Van Schaik PH, Probst AH (1959) Agron Jour *51,* 510

985. Greer HAL, Anderson IC (1965) Crop Sci *5,* 229
986. Cothren JT, Stutte CA (1973) Phyton *31,* 137
987. Stutte CA (1974) Soybean Production, Marketing and Use, TVA Bull Y–69, 72
988. Stutte CA (1974) Roy Soc New Zealand Bull. *12,* 923
989. Stutte CA, Cothren JT, Bryant SD (1975) Arkansas Farm Res *24* (4), 13
990. Cothren JT, Rutledge SR, Stutte CA (1975) Arkansas Farm Res *24* (3), 3
991. Monn DK, Son ER (1972) Jour Korean Soc Crop Sci *12,* 43
992. Long RC, Weybrew JA, Woltz WG, Dunn CA (1974) Tobacco Sci *18,* 73
993. Son ER (1974) Jour Korean Soc Crop Sci *15,* 145
994. Domir SC, Foy CL (1976) Tobacco Sci *20,* 158
995. Blatt CR, Sponagle AG (1977) Can Jour Plant Sci *57,* 1179
996. Walker EK (1977) Can Jour Plant Sci *57,* 819
997. Chakraborty MK, Prabhu SR, Suryanarayana TK (1979) Indian Jour Agri Sci *49,* 520
998. Robinson RW, Wilczynski H, Dennis FG, Bryan HH (1968) Proc Amer Soc Hort Sci *93,* 823
999. Sims WL, Fosse R, Ede L (1970) Vegetable Briefs *134,* 1
1000. Dostal HC, Wilcox GE (1971) Jour Amer Soc Hort Sci *96,* 656
1001. Splittstoesser WE, Vandemark JS (1971) Jour Amer Soc Hort Sci *96,* 564
1002. Saimbhi MS, Kanwar JS, Nandpuri KS (1974) Jour Res Punjab Agri Univ *12,* 128
1003. Saimbhi MS, Kanwar JS, Nandpuri KS (1975) Jour Res Punjab Agri Univ *13,* 286
1004. Bagar MR, Edwards RA, Lee TH (1975) Aust Jour Exp Agri Animal Husbandry *15,* 839
1005. Lukasik S (1977) Arch Gartenbau *25,* 61
1006. Hofer W, Schliebs R, Schmidt RR, Eue L: US 3 771 992 (23.9.70/13.11.73)
1007. Supriewska JH (1963) Bull Pol Acad Sci (Biol Sci) *11,* 165
1008. Dyson PW (1972) Jour Hort Sci *47,* 215
1009. Thomas TH, Currah IE, Salter PJ (1973) Ann Appl Biol *75,* 63
1010. von Kuhn H, Raafat A (1975) Z Pflanzenernährung Bodenkunde *4/5,* 505
1011. Weston GD, Thomas TH (1980) Jour Hort Sci *55,* 253
1012. Boe AA, Lee TS, Tapio DD, Banko TJ (1973) HortScience *8,* 497
1013. Abdalla IM, Helal RM, Zaki ME-S (1979) Ann Agri Sci (Moshtohor) *12,* 199
1014. Smock RM, Gross CR (1947) Proc Amer Soc Hort Sci *49,* 67
1015. Smock RM, Edgerton LJ, Hoffman MB (1952) Proc Amer Soc Hort Sci *60,* 184
1016. Baker DR, Hyzak DL: US 4 148 926 (4.10.77/10.4.79)
1017. Baker DR, Hyzak DL: US 4 148 927 (4.10.77/10.4.79)
1018. Baker DR, Hyzak DL: US 4 216 238 (13.2.79/5.8.80)
1019. Laibach F, Kribben FJ (1950) Naturwissenschaften *43,* 284
1020. Laibach F, Kribben FJ (1950) Ber Deutsch Bot Ges *62,* 53
1021. Heslop-Harrison J (1959) Jour Linnean Soc Bot (London) *56,* 269
1022. Heslop-Harrison J (1957) Biol Dev *32,* 38
1023. Galun E (1959) Phyton *13,* 1
1024. Prakash G (1977) Curr Sci (India) *46,* 328
1025. Peterson CE, Anhder LD (1960) Science *131,* 1673
1026. Peterson CE, Anhder LD (1960) Science *131,* 1673
1027. Bukovac MJ, Wittwer SH (1961) Adv Chem Ser *28,* 80
1028. Pike LM, Peterson CE (1969) Euphytica *18,* 106
1029. Krishnamoorthy HN (1972) Plant Cell Physiol *13,* 381
1030. Lower RL, Pharr DM, Horst EK (1978) Cucurbit Genetics Cooperative *1,* 8
1031. Tolla GE, Peterson CE (1979) HortScience *14,* 542
1032. McMurray AL, Miller CH (1968) Science *162,* 1396
1033. Iwahori S, Lyons JM, Sims WL (1969) Nature *222,* 271
1034. Rudich J, Halevy AH, Kedar N (1969) Planta *86,* 69
1035. Shannon S, de la Guardia MD (1969) Nature *223,* 186
1036. Robinson RW, Shannon S, de la Guardia MD (1969) BioScience *19,* 141
1037. McMurray AL, Miller CH (1969) Jour Amer Soc Hort Sci *94,* 400
1038. Freytag AH, Lira EP, Isleib DR (1970) HortScience *5,* 509
1039. Rudich J, Kedar N, Halevy AH (1970) Euphytica *19,* 47
1040. Coyne DP (1970) HortScience *5,* 227

1041. George WL (1971) Jour Amer Soc Hort Sci *96,* 152
1042. Robinson RW (1978) Cucurbit Genetics Cooperative *1,* 10
1043. Splittstoesser WE (1970) Physiol Plant *23,* 762
1044. Hopping ME, Hawthorne BT (1979) NZ Jour Exp Agri *7,* 399
1045. Baker EC, Bradley GA (1976) HortScience *11,* 140
1046. Sams CE, Krueger WA (1977) HortScience *12,* 162
1047. Bose TK, Nitsch JP (1970) Physiol Plant *23,* 1206
1048. Kubicki B (1969) Gen Polonica *10,* 145
1049. Karchi Z (1970) Jour Amer Soc Hort Sci *95,* 515
1050. Loy JB (1971) Jour Amer Soc Hort Sci *96,* 641
1051. Lee CW, Janick J (1978) HortScience *13,* 195
1052. Loy JB (1978) Cucurbit Genetics Cooperative *1,* 18
1053. Loy JB, Natti TA, Zak CD, Fritts SK (1979) Jour Amer Soc Hort Sci *104,* 100
1054. Mohan Ram HY, Jaiswal VS (1970) Experientia *26,* 214
1055. Mohan Ram HY, Jaiswal VS (1972) Planta *105,* 263
1056. Corley RHV (1976) Jour Exp Bot *27,* 553
1057. Mitchell WD, Wittwer SH (1962) Science *136,* 880
1058. Halevy AH, Rudich J (1967) Physiol Plant *20,* 1052
1059. Amruthavalli SA (1978) Curr Sci (India) *47,* 929
1060. Krishnamoorthy HN (1971) Ztschr Pflanzenphysiol *65,* 88
1061. Irving RM (1968) Okla Agri Exp Stn Bull T–128, 1
1062. Wittwer SH, Hillyer JG (1954) Science *120,* 893
1063. Chailakhyan MK, Khryanin VN (1979) Doklady Akad Nauk USSR *244,* 1037
1064. Takahashi H, Suge H, Saito T (1980) Plant Cell Physiol *21,* 525
1065. Patel B, Patel VJ (1979) Pesticides (India) *13,* 28
1066. Owens KW, Tolla GE, Peterson CE (1980) HortScience *15,* 256
1067. Byers RE, Baker LR, Sell HM, Herner RC, Dilley DR (1972) Proc Nat'l Acad Sci (USA) *69,* 717
1068. Atsmon D, Tabbak C (1979) Plant Cell Physiol *20,* 1547
1069. Owens KW, Peterson CE, Tolla GE (1980) HortScience *15,* 654
1070. Tolla GE, Peterson CE (1979) HortScience *19,* 542
1071. Nooden LD, Lindoo SJ (1978) What's New in Plant Physiol *9* (7), 25
1072. Gepstein S, Thimann KV (1980) Proc Nat'l Acad Sci (USA) *77,* 2050
1073. Thomas H, Stoddard JL (1980) Ann Rev Plant Physiol *31,* 83
1074. Osborne DJ (1967) Symp Soc Exp Biol *21,* 305
1075. Fletcher RA (1969) Planta *89,* 1
1076. Woolhouse HW (1974) Sci Prog Oxford *61,* 123
1077. Egharevba PN (1978) Cereal Res Comm *6,* 405
1078. Nooden LD, Kahanak GM, Okatan Y (1979) Science *206,* 841
1079. Lewis LN, Coggins CW, Labanauskas CK, Duggar WM (1967) Plant Cell Physiol *8,* 151
1080. Coggins CW, Eaks IL (1967) Calif Citrograph *52,* 475
1081. Coggins CW (1969) Proc 1st Internat'l Citrus Symp *3,* 1177
1082. Coggins CW, Scora RW, Lewis LN, Knapp JCF (1969) Jour Agri Food Chem *17,* 807
1083. Coggins CW (1972) Acta Hort *34,* 469
1084. Arguelles T, Gaurdiola JL (1977) Jour Hort Sci *52,* 199
1085. Wang CY (1977) HortScience *12,* 54
1086. Baker JE, Wang CY, Lieberman M, Hardenburg R (1977) HortScience *12,* 38
1087. Wang CY, Baker JE (1979) HortScience *14,* 59
1088. Wang CY, Baker JE, Hardenburg RE, Liebermann M (1977) Jour Amer Soc Hort Sci *102,* 517
1089. Dilley DR, Carpenter WJ (1975) Acta Hort *41,* 117
1090. Coggins CW, Hield HZ, Eaks IL, Lewis LN, Burns RM (1965) Calif Citrograph *50,* 457
1091. Riehl LA, Coggins CW, Carman GE (1965) Calif Citrograph *51,* 2
1092. Coggins CW, Henning GL, Atkin DR (1974) Jour Amer Soc Hort Sci *99,* 197
1093. El.-Zeftawi BM (1978) Sci Hort *12,* 177
1094. Kirby BW, Stelzer LR (1968) Proc Beltwide Cotton Prod Res Conf *22,* 68
1095. McMeans JL, Walhood VT, Carter LM (1966) Agron Jour *58,* 91

1096. Cathey GW (1979) Agron Jour *71,* 505
1097. Singh G, Kumar S (1978) Ind Jour Agri Sci *48,* 632
1098. Andrioli C, Ebeltoft DC (1979) 1st Ann Semin Nac Pesqui Soja (Brazil) 1978–2; 277
1099. Shaw WC (1978) Ann. Rpt. Program 22–677 Crop Prod. Efficiency Res., USDA, 6
1100. Whigham DK, Stoller EW (1979) Agron Jour *71,* 630
1101. Anonymous (1979) Agrichemical Age *23* (8), 32
1102. Smith RJ, Hinkle DA, Williams FJ (1959) Ark Agri Exp Stn Bull *619,* 1
1103. Tullis EC (1951) Proc So Weed Conf *4,* 1
1104. Hinkle DA (1954) Proc Rice Tech Working Group *6,* 16
1105. Hinkle DA (1952) Proc So Weed Conf *5,* 175
1106. Williams AH (1952) Down to Earth *8* (2), 2
1107. Hinkle DA (1953) Proc So Weed Conf *6,* 72
1108. Eastin EF (1978) Crop Sci *18,* 1068
1109. Eastin EF (1980) Crop Sci *20,* 389
1110. Ramanujam T, Nagarajan M, Sethuraman S, Pillaiyar P (1979) Pesticides (India) *13* (8), 40
1111. Chrominski A, Rozej B (1976) Potato Res *19,* 123
1112. Ames RB, Brewer AD, McIntire WS (1974) Proc Beltwide Cotton Prod Res Conf *28,* 61
1113. Bohne PW (1977) Proc Plant Growth Reg Work Group *4,* 172
1114. Ames RB, Corkins JP (1978) Proc Plant Growth Reg Work Group *5,* 132
1115. Kochmann W, Naumann K, Roethling T: German (DDR) Pat 140 405 (1.11.78/5.3.80)
1116. Bovey RW, McCarty MK (1965) Crop Sci *5,* 523
1117. Bovey RW, Miller FR, Baur JR (1975) Agron Jour *67,* 618
1118. Anonymous (1980) Agrichemical Age *24* (8), 42
1119. Anonymous (1980) Pest Toxic Chemical News *8* (40), 20
1120. Selevicius A, Jundulas J (1980) Khim Sel'sk Khoz *18,* 45
1121. Hoffman OL (1953) Plant Physiol *28,* 622
1122. Hoffman OL (1962) Weeds *10,* 322
1123. Hoffman OL, Gull PW, Zeisig HC, Epperly JR (1960) Proc North Central Weed Control Conf *17,* 20
1124. Hoffman OL (1978) Chemistry and Action of Herbicide Antidotes. In: Pallos FM, Casida JE, (eds) Academic Press, New York
1125. Hoffman OL: US 3 564 768 (25.10.68/23.2.71)
1126. Stephenson GR, Chang FY (1978) Chemistry and Action of Herbicide Antidotes. In: Pallos FM, Casida JE (eds) Academic Press, New York
1127. Slife FW (1978) Chemistry and Action of Herbicide Antidotes. In: Pallos FM, Casida JE (eds) Academic Press, New York
1128. Pallos FM, Brokke ME, Arneklev DR: US 4 021 224 (7.12.75/3.5.77)
1129. Pallos FM, Gray RA, Arneklev DR, Brokke ME (1978) Chemistry and Action of Herbicide Antidotes. In: Pallos FM, Casida JE (eds) Academic Press, New York
1130. Leavitt JRC, Penner D (1978) Weed Sci *26,* 653
1131. Gray RA, Joo GK (1978) Chemistry and Action of Herbicide Antidotes. In: Pallos FM, Casida JE (eds) Academic Press, New York
1132. Miaullis JB, Thomas VM, Gray RA, Murphy JJ, Hollingworth RM (1978) Chemistry and Action of Herbicide Antidotes. In: Pallos FM, Casida JE (eds) Academic Press, New York
1133. Phillips BA, Bhagsari AS (1978) Chemistry and Action of Herbicide Antidotes. In: Pallos FM, Casida JE (eds) Academic Press, New York
1134. Fujimoto TT: US 4 009 022 (30.10.75/22.2.77)
1135. Ellis JF, Peek JW, Boehle J, Muller G (1980) Weed Sci *28,* 1
1136. Nyffeler A, Gerber HR, Hensley JR (1980) Weed Sci *28,* 6
1137. Miller SD, Nalewaja JD (1980) Agron Jour *72,* 662
1138. Houbion JA, Schafer DE: US 4 182 622 (28.6.78/8.1.80)
1139. Pallos FM, Arneklev DR: US 4 197 111 (27.9.78/8.4.80)
1140. Martin H: US 4 220 464 (30.8.78/2.9.80)
1141. Martin H: US 4 225 334 (26.3.79/30.9.80)
1142. Martin H: US 4 227 917 (26.3.79/14.10.80)

1143. Peterson LW: US 4 230 482 (21.5.79/28.10.80)
1144. Czajkowski AJ, Schafer DE: US 4 231 786 (26.12.74/4.11.80)
1145. Peterson LW: US 4 231 785 (21.5.79/4.11.80)
1146. Hawkins AF, Lewis T, Jones I: US 4 242 121 (8.12.78/30.12.80)
1147. Casida JE (1978) Chemistry and Action of Herbicide Antidotes. In: Pallos FM, Casida JE (eds) Academic Press, New York
1148. Okii M, Matsukuma I, Konnai M, Takematsu T (1979) Zasso Kenkyu *24,* 10
1149. Binning LKD, Penner D, Meggit WF (1971) Weed Sci *19,* 73
1150. Carson AG, Bandeen JD (1975) Can Jour Plant Sci *55,* 795
1151. Baradari MR, Haderlie LC, Wilson RG (1980) Weed Sci *28,* 197
1152. Anderson KJ, Leighty EG, Takahashi MT (1972) Jour Agri Food Chem *20,* 649
1153. Northrop JH (1963) Jour Gen Physiol *46,* 971
1154. Environmental Protection Agency (1977) Federal Register *42* (208), 56920
1155. Haley TJ (1977) Jour Toxicol Environ Health *2,* 1085
1156. Dickens R, Jones HEH (1965) Brit Jour Cancer *19,* 392
1157. Epstein SS, Mantel N (1968) Internat'l Jour Cancer *3,* 325
1158. Das RK, Manna CK (1969) Proc 56th Ind Sci Cong *111,* 453
1159. Oku K (1977) Senshokutai *5,* 138
1160. Kihlman BA, Anderson HC (1980) Environ Exp Bot *20,* 271
1161. Hartley-Asp B, Anderson HC, Sturelid S, Kihlman BA (1980) Environ Exp Bot *20,* 119
1162. Biswas PK, Hall O, Mayberry BD (1967) Physiol Plant *20,* 119
1163. Fischnich O, Patzold C, Schiller C (1958) Eur Potato Jour *1,* 25
1164. Greulach VA, McKenzie J, Stacy EM (1951) Biol Bull *101,* 285
1165. Akin FJ, Norred WP (1978) Toxicol Appl Pharmacol *43,* 287
1166. Nishi Y, Mori M, Inui N (1979) Mutation Res *67,* 249
1167. Gehring PJ, Betso JE (1978) Ecol Bull (Stockholm) *27,* 122
1168. Hansen WH, Quaife ML, Habermann RT, Fitzhugh OG (1971) Toxicol Appl Pharmacol *20,* 122
1169. Innes JRM, Ulland BM, Valerio MG, Petrucelli L, Fishbein L, Hart ER, Pallotta AJ, Bates RR, Falk HL, Gart JJ, Klein M, Mitchell I, Peters J (1969) Jour Nat'l Cancer Inst *42,* 1101
1170. Anonymous (1972) FAO/WHO, WHO Tech Rep Ser No 502, FAO Agri Studies, No *88,* 1
1171. Jenssen D, Renberg L (1976) Chem-Biol Interactions *14,* 291
1172. Elo H, Ylitalo P (1977) Acta Pharmacol Toxicol *41,* 280
1173. Elo H, Ylitalo P (1979) Toxicol Appl Pharmacol *51,* 439
1174. Burchfield HP, Stores EE (1977) Adv Modern Toxicol *3,* 319
1175. Theiss JC, Shimkin MB (1980) Food Cosmetic Toxicol *18,* 129
1176. Blinn RC (1967) Jour Agri Food Chem *15,* 984
1177. Abou-Donia MB, Graham DG, Abdo KM, Komeil AA (1979) Toxicol *14,* 229
1178. Abou-Donia MB, Graham DG, Timmons PR, Reichert BL (1979) Neurotoxicol *1,* 425
1179. Abou-Donia MB, Graham DG, Timmons PR, Reichert BL (1980) Toxicol Appl Pharmacol *53,* 439
1180. Dasta JF (1980) Clin Toxicol Consultant *2,* 11
1181. Haley TJ (1979) Clin Toxicol *14,* 1
1182. Autor AP (1977) Biochemical Mechanisms of Paraquat Toxicity. Academic Press, New York
1183. Fairshter RD (1978) West Jour Med *128,* 56
1184. Bullivant CM (1966) Brit Med Jour *1,* 1272
1185. Smith RJ (1978) Science *200,* 417
1186. Manzo L, Gregotti C, Di Nucci A, Richelmi P (1979) Vet Human Toxicol *21,* 404
1187. John JA, Blogg CD, Murray FJ, Schwetz BA, Gehring PJ (1979) Teratology *19,* 321
1188. Gurd MR, Harmer GLM, Lessel B (1965) Food Cosmet Toxicol *3,* 883
1189. Verschuuren HG, Kroes R, den Tonkelaar EM (1975) Toxicol *3,* 349
1190. Hattula ML, Elo H, Reunanen H, Arstila AU, Sorvari TE (1977) Bull Environ Contam Toxicol *18,* 152

1191. Verschuuren HG, Kroes R, den Tonkelaar EM, van Esch GJ, Helleman PW (1976) Toxicol *5,* 371
1192. Anonymous (1980) Pest Toxic Chem News *9*(2), 11
1193. Peterson GE (1967) Agri Hist *41,* 243
1194. Pokorny R (1949) Jour Amer Chem Soc *63,* 1768
1195. Zimmerman PW, Hitchcock AE (1942) Contrib Boyce Thompson Inst *12,* 321
1196. Hammer CL, Tukey HB (1944) Science *100,* 154
1197. Marth PC, Mitchell JW (1944) Bot Gaz *106,* 224
1198. Hammer CL, Tukey HB (1944) Bot Gaz *106,* 232
1199. Johnson RR (1980) Crops Soil Mag *32* (9), 9
1200. Anonymous (1975) Farm Chem *138* (3), 15
1201. Hardy RWF (1979) Plant Regulation and World Agriculture. In: Scott TK (ed) Plenum Press, New York

Author Index

Aagensen, G. J. 34, 35, 134
Aalders, L. E. 74, 142
Abdalla, I. M. 82, 145
Abdalla, K. M. 6, 127
Abdel-Rahman, M. 76, 143
Abdo, K. M. 105, 148
Abeles, F. B. 8, 128
Abou-Donia, M. B. 105, 148
Abraham, P. D. 70, 141
Adams, J. B. 51, 136
Addison, B. M. 33, 133
Adenikinju, S. A. 27, 132
Adkisson, P. L. 52, 136, 137
Adriansen, E. 47, 135
Agabbio, M. 12, 128
Ahlrichs, L. E. 68, 141
Akin, F. J. 105, 148
Alamu, S. 13, 129
Aldrich, R. J. 35, 134
Aldwinckle, H. S. 53, 137
Alexander, A. G. 63, 64, 66, 67, 140
Alexander, M. W. 79, 144
Allard, H. A. 9, 128
Alley, C. J. 4, 126
Alphin, J. G. 82, 139
Al'sing, T. K. 16, 17, 130
Alves, A. S. 63, 64, 139, 140
Amen, R. D. 6, 127
Ames, R. B. 35, 95, 134, 147
Amruthavalli, S. A. 89, 146
Anderson, A. S. 5, 127
Anderson, H. C. 104, 148
Anderson, I. C. 81, 145
Anderson, K. J. 104, 148
Anderson, M. R. 34, 35, 134
Andreis, H. J. 63, 64, 67, 140
Andrioli, C. 94, 97, 147
Andris, H. 61, 80, 139, 144
Anhder, L. D. 89, 145
Antognozzi, E. 25, 131
Apte, S. S. 9, 128
Arendt, F. 20, 130
Arguelles, T. 91, 146
Arle, H. F. 21, 53, 130, 137
Arneklev, D. R. 99, 102, 147
Arnold, G. M. 14, 129
Arrington, E. H. 25, 132
Arstila, A. U. 106, 148
Arumagan, R. 17, 130
Arzee, T. 47, 135
Ashkar, S. A. K. 46, 135
Asokan, S. 63, 64, 139
Atkin, D. R. 92, 146
Atkins, H. 61, 139
Atkinson, W. O. 46, 135
Atsmon, D. 57, 90, 138, 146
Auchter, E. C. 24, 131
Audus, L. J. 4, 126
Austin, M. E. 74, 143
Austin, R. B. 6, 127
Autor, A. P. 105, 148
Avery, G. S. 4, 126
Azzi, G. M. 63, 64, 139, 140

Badrul Alam, A. F. M. 4, 126
Bagar, M. R. 82, 145
Bahn, G. W. 16, 129
Bailiss, K. W. 53, 137
Baker, D. R. 83, 145
Baker, E. C. 89, 146
Baker, J. E. 91, 146
Baker, L. R. 90, 146
Baker, W. R. 79, 144
Baldini, E. 49, 136
Balobin, V. N. 4, 126
Bandeen, J. D. 103, 148
Baneviciene, Z. 54, 137
Bangerth, F. 30, 56, 133, 138
Banko, T. J. 82, 145
Baptiste, E. D. C. 70, 141
Baradari, M. R. 103, 148
Barden, J. A. 25, 131
Bariola, A. L. 21, 53, 130, 137
Barreiro, M. 80, 144
Barritt, B. H. 29, 132
Barshi, G. 22, 131
Bartolini, G. 4, 126
Bartram, R. D. 72, 142
Basabe, P. J. 76, 143
Bastdorff, R. S. 29, 132
Basu, R. N. 5, 127
Batch, J. J. 15, 129
Bateman, G. L. 54, 137
Bates, J. F. 63, 140
Bates, R. R. 105, 148
Batjer, L. P. 24, 61, 72, 75, 131, 139, 142, 143
Bauman, A. J. 71, 141
Bauman, R. W. 79, 144
Baumgardner, R. A. 73, 142
Baur, J. R. 95, 97, 147
Beevers, L. 5, 127
Behnke-Rogers, J. 4, 126
Behrendt, S. 34, 134
Bellini, F. 4, 126
Bellman, S. K. 17, 130
Ben-Arie, R. 73, 142
Benedicto, F. 10, 128
Bennett, J. P. 4, 126
Bennett, M. D. 16, 129
Benoit, F. 48, 136
Ben-Tal, Y. 22, 131
Berrie, A. M. M. 6, 127
Bethell, R. 9, 128
Betso, J. E. 105, 148
Bhagsari, A. S. 99, 101, 147
Bhardwaj, S. N. 30, 132
Bhatnagar, P. S. 16, 129
Bhutani, V. P. 25, 132
Biddington, N. L. 6, 127
Bierhuizen, J. F. 55, 56, 137
Bieske, G. C. 63, 64, 139
Biggs, R. H. 79, 144
Bilgen, T. 16, 129
Binning, L. K. 61, 103, 139, 148

Birecka, H. 57, 138
Bishop, J. C. 7, 127
Biswas, P. K. 104, 148
Bittenbender, H. C. 33, 74, 133, 143
Blatt, C. R. 82, 145
Blazich, F. A. 5, 127
Bleinholder, H. 34, 134
Blencowe, J. W. 70, 141
Blinn, R. C. 105, 148
Blodgett, E. C. 80, 144
Blogg, C. D. 106, 148
Blumenfeld, A. 55, 57, 131, 138
Bocion, P. F. 30, 34, 35, 47, 57, 64, 82, 133–136
Bodden, J. J. 16, 129
Boe, A. A. 82, 145
Boehle, J. 99, 101, 147
Bohne, P. W. 95, 147
Bollinger, F. J. 17, 130
Borger, G. A. 55, 56, 138
Bose, T. K. 4, 5, 89, 126, 127, 146
Bosland, J. M. 77, 143
Bosshard, R. 65, 140
Bottle, R. T. 13, 129
Bouwkamp, J. C. 33, 133
Bovey, R. W. 95, 97, 147
Bradley, G. A. 89, 146
Bramblett, J. 76, 143
Bramlage, W. J. 61, 72, 73, 139, 142
Breece, J. R. 38, 47, 135
Brendel, T. P. 20, 105, 130
Brewer, A. D. 95, 147
Brian, P. W. 9, 128
Bridge, K. 70, 141
Briskovich, M. A. 34–36, 134
Brokke, M. E. 99, 102, 147
Brooks, B. M. 73, 142
Broomfield, J. 72, 142
Brown, C. L. 72, 142
Brown, C. M. 16, 129
Brown, H. M. 76, 143
Brown, L. C. 11, 128
Brown, R. H. 79, 144
Brown, S. A. 25, 132
Browning, G. B. 27, 76, 132, 143
Bruzzone, E. 9, 128
Bryan, H. H. 82, 145
Bryant, S. D. 82, 145
Buchanan, D. W. 79, 144
Buchenauer, H. 54, 137
Buckman, S. J. 67, 141
Buescher, R. W. 61, 138
Bukovac, M. J. 6, 7, 24, 27, 89, 127, 130–132, 145
Bullivant, C. M. 106, 148
Bullock, R. 76, 143
Burchfield, H. P. 105, 148
Burg, E. A. 73, 142
Burg, S. P. 73, 142
Burkholder, C. L. 24, 131
Burns, R. M. 6, 25, 91, 92, 127, 132, 146
Burton, C. L. 74, 143
Bush, L. P. 35, 134
Bushong, J. W. 35, 65, 134
Bustamante, M. 30, 133
Buta, J. G. 46, 135
Buxton, J. 48, 136
Byers, R. E. 24, 25, 73, 78, 79, 90, 131, 142, 144, 146
Bynum, E. K. 51, 136

Caceres, J. R. 56, 138
Campbell, J. M. 52, 136
Cannell, M. G. R. 27, 76, 132, 143
Cantliffe, D. J. 17, 29, 61, 130, 132, 138
Cappelini, P. 61, 139
Carlisle, D. B. 52, 136
Carlson, A. E. 65, 140
Carlson, P. 33, 133
Carlson, S. P. 22
Carman, G. E. 58, 92, 138, 146
Carnill, G. L. 25, 132
Carns, H. R. 20, 130
Carolus, R. L. 80, 144
Carpenter, W. J. 91, 146
Carson, A. G. 103, 148
Carter, G. H. 80, 144
Carter, L. N. 94, 146
Carter, O. G. 6, 127
Casida, J. E. 100, 148
Cassells, A. C. 53, 137
Castro, P. R. C. 59, 138
Cathey, G. W. 11, 19, 21, 53, 93, 97, 128, 130, 137, 147
Cathey, H. M. 37, 47, 57, 135, 138
Cawthon, D. L. 74, 142
Chacko, E. K. 70, 141
Chacravarti, A. S. 63–67, 139
Chailakhyan, M. K. 90, 146
Chakraborty, M. K. 82, 145
Chakrawar, V. R. 78, 143
Chamberlain, J. D. 37, 74, 135, 142
Chandrasekera, L. B. 70, 141
Chang, F. Y. 99, 101, 102, 147
Chang, W. C. 50, 136
Chapman, G. W. 70, 141
Chapman, L. S. 63, 64, 67, 139
Chappell, W. E. 35, 134
Charlton, J. L. 33, 133
Charpentier, L. J. 51, 136
Chauhan, S. V. S. 17, 130
Chen, I. Y. 22, 46, 131, 135
Chiba, K. 24, 131
Child, R. D. 61, 139
Chin, C. 4, 126
Chin, T. Y. 5, 127
Ching, T. T. 77, 143
Chittiraichelvan, T. 70, 141
Chopra, V. L. 17, 130
Choudhurry, A. R. 16, 130
Chowdhurry, I. R. 33, 133
Christensen, J. H. 26, 132
Christodoulou, A. 78, 143
Chrominski, A. 33, 95, 97, 133, 147
Chua, S. E. 70, 141
Chundawat, B. S. 78, 144
Cibes, H. R. 9, 128
Cibulsky, R. J. 30, 133
Clark, F. 46, 135
Clark, H. E. 9, 128
Clark, R. K. 21, 130
Clerc, P. 6, 127
Cleveland, T. C. 53, 137
Clore, W. J. 27, 78, 132, 144
Clowes, M. S. J. 63, 64, 67, 68, 76, 139, 143
Coal, D. F. 6, 127
Coartney, J. S. 35, 134
Coats, G. E. 6, 127
Cocker, F. M. 53, 137
Coggins, C. W. 6, 25, 58, 60, 61, 75, 76, 91, 92, 127, 132, 138, 143, 146
Coleman, W. 5, 127
Collins, D. J. 46, 135
Collins, K. 76, 143
Colombo, R. 6, 127
Cook, C. E. 46, 135
Cooke, A. R. 8, 16, 46, 57, 128, 129, 135
Coombe, B. G. 6, 29, 78, 127, 132, 143

Cooper, P. E. 74, 142
Cooper, W. C. 9, 21, 128, 130
Coorts, G. D. 35, 134
Copping, L. G. 66, 141
Coppock, G. E. 21, 130
Cordukes, W. E. 34, 35, 134
Corgan, J. N. 9, 128
Corkins, J. P. 95, 147
Corley, R. H. V. 89, 146
Cormack, N. R. 37, 135
Cothren, J. T. 34, 82, 134, 145
Coupe, M. 70, 141
Coyne, D. P. 89, 145
Craig, D. L. 74, 142, 143
Crandall, P. C. 37, 74, 135, 142
Crane, J. C. 9, 73, 128, 142
Crews, E. R. 72, 142
Criley, R. A. 4, 126
Crowe, A. D. 72, 73, 142
Crowe, G. B. 20, 130
Currah, I. E. 82, 145
Currier, H. G. 35, 134
Czajkowski, A. J. 100–102, 148

Daft, M. J. 53, 137
Dallyn, S. L. 7, 127
D'Amico, J. J. 17, 130
Dana, M. N. 61, 139
Daniell, J. W. 25, 131
Das, N. 9, 128
Das, P. 4, 126
Das, R. C. 4, 126
Das, R. K. 104, 148
Das, V. S. R. 55, 56, 138
Das Gupta, D. K. 79, 144
Dasta, J. F. 105, 106, 148
Daughtry, C. S. 79, 144
d'Auzac, J. 70, 141
Davenport, D. C. 55, 56, 138
David, M. 25, 132
Davidson, J. H. 24, 131
Davies, F. S. 21, 130
Davies, F. T. 4, 126
Davies, W. J. 55, 56, 137, 138
Davis, D. 53, 137
DeBenedict, C. 60, 138
deBie, J. 47, 136
Dedolph, R. R. 73, 142
Deidda, P. 12, 128
de Jonge, P. 70, 141
Dekazos, E. D. 74, 143
de la Guardia, M. 29, 89, 132, 145
Delegher-Langohr, V. 5, 127
della Pieta, S. 22, 131
de Mur, A. 30, 35, 57, 64, 82, 133, 134
Dennis, F. G. 9, 82, 128, 145
den Tonkelaar, E. M. 106, 148, 149
de Sacks, R. L. 16, 130
Desai, B. B. 73, 142
Deshpande, P. B. 73, 142
de Silva, W. H. 30, 34, 35, 47, 57, 64, 82, 133–136
De Stefano, R. P. 63, 64, 67, 69, 140, 141
Devlin, R. M. 59, 61, 138, 139
Dhuria, H. A. 25, 132
Dickens, R. 104, 148
Dickenson, P. B. 70, 141
Dicks, J. W. 34, 134
Dike, R. B. 4, 126
Dilley, D. R. 33, 80, 90, 91, 133, 144, 146
Dimetry, N. Z. 52, 136
Dimond, A. E. 53, 137
DiNucci, A. 106, 148
Dirou, J. F. 81, 144
Doherty, J. H. 61, 138
Dolgopolova, L. N. 55, 137
Domir, S. C. 62, 82, 139, 145
Donno, G. 22, 131
Dostal, H. C. 73, 78, 82, 142, 144, 145
Drake, S. R. 61, 139
Dressel, J. 34, 134
Drew, J. 72, 142
Dua, I. S. 30, 132
Dudley, J. W. 16, 130
Duggar, W. M. 91, 146
Duich, J. M. 34, 35, 134
Dunberg, A. 14, 129
Dunn, C. A. 82, 145
Dutky, S. R. 33, 133
Dutton, W. C. 24, 131
Dyck, R. 80, 144
Dyson, P. W. 82, 145

Eaks, H. 75, 91, 92, 143, 146
Earley, E. B. 16, 129
Eastin, E. F. 94, 97, 98, 147
Eastwood, D. 50, 136
Eaton, F. W. 16, 130
Ebeltoft, D. C. 94, 97, 147
Eck, P. 61, 74, 139, 143
Ede, L. 82, 145
Edgerton, L. J. 9, 24, 25, 72, 78, 83, 128, 131, 132, 142, 144, 145
Edwards, R. A. 82, 145
Eenink, A. H. 17, 130
Eggenberg, P. 30, 35, 57, 64, 82, 133, 134
Egharevba, P. N. 91, 146
Ehrenfreund, J. 35, 36, 134
Eichmeier, J. 52, 136
Eidt, P. C. 55, 137
El-Deen, S. A. S. 4, 126
El-Hamady, M. 22, 131
El-Ibrashy, M. T. 52, 136
Elkins, D. M. 34–36, 134
Ellis, J. F. 99, 101, 147
Ellis, M. T. 21, 130
Ellis, P. E. 52, 136
El-Mahdy, A. 4, 126
El-Masiry, H. H. 6, 127
El-Motaz-Bellah, M. 13, 129
Elo, H. 105, 106, 148
Else, J. A. 78, 143
El-Shafie, S. A. 13, 129
El-Wakeel, A. T. 6, 127
Ely, D. G. 35, 134
El-Zeftawi, B. M. 92, 146
Embree, C. G. 72, 73, 142
Emerson, F. H. 24, 73, 78, 79, 131, 142, 144
Engel, R. E. 35, 134
Engelhaupt, M. E. 46, 135
England, D. J. F. 49, 136
Enos, H. I. 72, 142
Enright, L. J. 4, 127
Enzie, J. V. 24, 131
Epperly, J. R. 99, 100, 147
Epstein, S. S. 104, 148
Erafa, A. E. 13, 129
Erez, A. 37, 135
Erkan, Z. 56, 138
Erner, Y. 76, 143
Erwin, D. C. 54, 137
Erxleben, H. 3, 126
Escobar, D. E. 34, 134
Escober, T. R. 63, 140
Ethredge, W. J. 79, 144
Eue, L. 82, 145
Evans, A. W. 64, 140
Evans, H. 63, 140

Evans, H. R. 9, 128
Evans, S. D. 76, 143
Evenari, M. 6, 127
Eynard, I. 27, 132

Fadl, M. S. 4, 22, 126, 131
Fairey, D. T. 16, 18, 129
Fairshter, R. D. 106, 148
Falk, H. L. 105, 148
Faust, M. 72, 142
Fay, R. D. 27, 132
Fedin, M. A. 16, 17, 130
Fedurko, T. 4, 126
Felauer, E. E. 35, 134
Ferrara, E. 22, 131
Ferre, D. C. 30, 73, 133, 142
Fewkes, D. W. 63, 139
Fischnich, O. 104, 148
Fishbein, L. 105, 148
Fisher, J. B. 14, 129
Fitzhugh, O. G. 105, 148
Flemming, H. K. 29, 132
Fletcher, A. M. 14, 129
Fletcher, R. A. 53, 91, 137, 146
Flippen-Anderson, J. L. 33, 133
Florencio, A. C. 50, 136
Florido, L. V. 4, 127
Fogle, H. W. 25, 80, 132, 144
Ford, J. H. 76, 143
Ford, Z. T. 82, 139
Forsythe, F. R. 72–74, 142, 143
Fosse, R. 82, 145
Foster, C. A. 16, 130
Foy, C. L. 33, 62, 82, 133, 139, 145
Frans, R. F. 11, 128
Franz, J. E. 67, 141
Fraser, R. S. S. 53, 137
Freeman, B. 22, 130
Freeman, T. G. 55, 137
Frenkel, C. 80, 144
Freytag, A. H. 89, 90, 145
Fridinger, T. L. 35, 134
Frieburg, S. R. 73, 142
Fritts, S. K. 89, 90, 146
Fritz, W. 33, 133
Fujimoto, T. T. 99, 101, 147
Fukuhara, T. 46, 135
Fulk-Bringman, S. S. 33, 76, 133, 143
Funt, R. C. 73, 142
Furuta, T. 38, 47, 135
Gaertner, V. R. 65, 140
Gagneja, M. R. 16, 130
Galanopoulou, S. 16, 129
Gale, J. 55, 56, 138
Galena, F. E. 21, 130
Gallasch, P. T. 25, 132
Galun, E. 89, 145
Gambrell, C. E. 25, 73, 78, 131, 132, 142, 144
Gamez, H. 56, 138
Gandia, H. 9, 128
Gane, R. 73, 142
Garcia, G. 63, 64, 140
Garner, W. W. 9, 128
Garrod, J. F. 66, 141
Gart, J. J. 105, 148
Garth, J. K. L. 37, 135
Garza, M. V. 34, 134
Gates, D. W. 35, 65, 134
Gaurdiola, J. L. 91, 146
Gausman, H. W. 34, 134
Gehring, P. J. 105, 106, 148
Geiszler, J. 25, 132
George, E. F. 66, 141
George, M. J. 70, 141
George, W. L. 89, 146
Gepstein, S. 91, 146
Gerber, H. R. 99, 101, 147
Gerdts, M. H. 81, 144
Gerin, G. 25, 131
Ghani, A. 70, 141
Gibeault, V. A. 34, 35, 134
Gil, G. 9, 75, 128, 143
Ginoza, H. 32, 133
Gitler, C. 57, 138
Giulivo, C. 25, 131
Glenn, S. 35, 134
Godley, G. L. 30, 133
Godoy, O. P. 50, 136
Goldin, E. 79, 144
Goldschmidt, E. E. 76, 143
Gomez, E. 9, 128
Gomez, J. B. 70, 141
Gonzales, M. Y. 63, 64, 67, 140
Goodwin, P. B. 27, 61, 132, 138
Goolambossen, M. 63, 64, 139
Gorbet, D. W. 79, 144
Gordon, R. S. 78, 144
Gorecki, R. S. 4, 126
Goren, R. 13, 76, 129, 143
Gorini, F. L. 61, 139
Goss, R. L. 35, 134
Goto, S. 73, 142
Goveas, J. 10, 128
Gowing, D. P. 8, 128
Graf, H. R. 47, 136
Graham, B. A. 35, 134
Graham, D. G. 105, 148
Granger, R. L. 57, 74, 138, 142
Grauslund, J. 72, 142
Gray, R. A. 99, 102, 147
Green, N. A. 33, 133
Greene, D. W. 61, 72, 73, 139, 142
Greene, G. M. 30, 133
Greenhalgh, W. J. J. 24, 25, 30, 78, 131, 133, 144
Greenwood, M. S. 14, 129
Greer, H. A. L. 81, 145
Gregotti, C. 106, 148
Gressel, J. 47, 135
Greulach, V. A. 104, 148
Greyson, R. I. 5, 127
Griggs, W. 9, 128
Gritton, E. T. 34, 133
Gross, C. R. 83, 145
Grove, M. D. 33, 133
Gruzdev, L. G. 59, 138
Guelfat-Reich, S. 73, 142
Gull, P. W. 99, 100, 147
Gurd, M. R. 106, 148
Guyer, G. 52, 136
Guyot, A. 9, 128
Guyot, H. M. 67, 141
Gyska, M. N. 16, 17, 130

Haagen-Smit, A. J. 3, 126
Haard, N. F. 80, 144
Habermann, R. T. 105, 148
Hackett, W. P. 47, 135
Hacskaylo, J. 11, 128
Haderlie, L. C. 103, 148
Hagan, R. M. 55, 56, 138
Hale, C. R. 78, 143
Halevy, A. H. 13, 52, 55, 89, 129, 136, 137, 145, 146
Haley, T. J. 104–106, 148
Hall, A. E. 61, 138
Hall, C. B. 79, 144
Hall, O. 104, 148
Hallock, D. L. 79, 144
Halloin, J. M. 6, 127
Halsey, D. 77, 143
Hamilton, R. A. 73, 142
Hamm, P. C. 35, 64, 65, 134, 140
Hammer, C. L. 107, 149
Hammer, O. H. 24, 131

Hammett, L. K. 61, 139
Hangarter, R. 33, 133
Hanisch, B. 20, 130
Hansen, E. 80, 144
Hansen, D. J. 17, 130
Hansen, W. H. 105, 148
Harbaugh, B. K. 13, 34, 48, 129, 134, 136
Hardenburg, R. 91, 146
Hardy, R. W. F. 109, 149
Harley, C. P. 24, 131
Harmer, G. L. M. 106, 148
Harner, R. C. 61, 138
Harrington, T. A. 72, 142
Hart, E. R. 105, 148
Hartley-Asp, B. 104, 148
Hartmann, H. T. 4, 22, 126, 131
Hartung, W. 5, 127
Hartzook, A. 79, 144
Harwood, R. F. 51, 136
Hashizume, H. 14, 129
Haskal, A. 22, 131
Hatley, O. E. 76, 143
Hattula, M. L. 106, 148
Havis, L. 78, 144
Hawker, J. S. 78, 143
Hawkins, A. F. 100, 148
Hawthorne, B. T. 89, 146
Hayman, E. P. 60, 71, 138, 141
Haynes, R. L. 80, 144
Hecker, R. J. 16, 129
Hedberg, P. R. 27, 132
Helal, R. M. 82, 145
Helleman, P. W. 106, 149
Helseth, N. T. 46, 135
Hemstreet, S. 34, 35, 134
Hendershott, C. H. 21, 130
Henley, R. W. 38, 135
Henneberry, T. J. 53, 137
Henning, G. L. 60, 61, 92, 138, 146
Henry, R. J. 13, 129
Hensley, J. R. 46, 99, 101, 135, 147
Hermann, L. 76, 143
Herner, R. C. 80, 90, 144, 146
Herrara-Aguirre, E. 24, 131
Herzog, G. A. 11, 128
Heslop, A. J. 22, 131
Heslop-Harrison, J. 89, 145
Hess, P. J. 80, 144
Heuser, C. W. 5, 127
Hicks, D. R. 76, 143
Hicks, J. R. 9, 128
Hieke, K. 4, 126
Hield, H. Z. 25, 34, 35, 47, 75, 76, 91, 92, 132, 134–136, 143, 146
Hilliard, B. G. 47, 135
Hillyer, J. G. 89, 146
Hinkle, D. A. 94, 97, 98, 147
Hiron, R. W. P. 55, 137
Hirose, K. 25, 132
Hitchcock, A. E. 3, 107, 126, 149
Hittle, C. N. 16, 130
Hlavic, M. 16, 130
Ho, F. W. 10, 63–67, 128, 140
Hockett, E. A. 16, 129
Hofer, W. 82, 145
Hoffman, M. B. 83, 145
Hoffman, O. L. 99–102, 147
Hogue, E. J. 57, 74, 138, 142
Holder, D. G. 63, 64, 69, 140, 141
Hollingworth, R. M. 99, 147
Holm, R. E. 21, 22, 130, 131
Honeyborne, C. H. B. 52, 136
Honeyborne, G. E. 61, 139
Hopping, M. E. 89, 146
Horsfall, F. 24, 25, 131, 132
Horst, E. K. 89, 145
Houbion, J. A. 100, 101, 147
Howell, G. S. 27, 74, 132, 143
Howell, S. L. 35, 134
Hoyle, B. J. 7, 127
Hoysler, V. 5, 127
Hsiao, A. I. 6, 127
Hsu, W. J. 60, 61, 71, 138, 141
Hughes, D. L. 77, 143
Hughes, J. L. 16, 17, 129
Hughes, W. G. 16, 129
Huglin, P. 37, 134
Hull, J. 27, 132
Humbert, R. P. 10, 63, 128, 139
Humphrey, W. 38, 135
Humphries, E. C. 33, 34, 133
Hunter, N. R. 33, 133
Hüppi, G. A. 34, 47, 134, 135
Hurney, A. P. 63, 64, 67, 139
Hutchinson, J. F. 24, 131
Hyzak, D. L. 83, 145
Ikonnikow, N. S. 17, 130
Ikuma, H. 6, 127
Ingalsbe, D. W. 80, 144
Ingram, J. W. 51, 136
Innes, J. R. M. 105, 148
Inui, N. 105, 148
Irving, M. R. 57, 138
Irving, R. M. 89, 146
Isikawa, H. 4, 127
Isleib, D. R. 89, 90, 145
Iwagaki, I. 25, 132
Iwahori, S. 25, 89, 90, 132, 145
Iwakiri, G. 9, 128

Ja'afar, H. 4, 126
Jacobson, M. 33, 133
Jaffar, H. 70, 141
Jagschitz, J. A. 34, 134
Jain, S. K. 17, 130
Jaiswal, V. S. 89, 146
Jan, C. C. 16, 17, 129, 130
Janick, J. 89, 146
Jaramillo, H. 63, 64, 140
Jaworski, E. G. 65, 140
Jeffcoat, B. 50, 136
Jensen, F. L. 61, 78, 139, 143
Jenssen, D. 105, 148
Jeppson, R. G. 76, 143
Jernberg, D. C. 28, 132
Jewiss, O. R. 50, 136
Johanson, S. 48, 136
John, J. A. 106, 148
Johnson, A. G. 34, 47, 134, 136
Johnson, D. S. 61, 139
Johnson, E. F. 4, 126
Johnson, J. O. 26, 132
Johnson, R. R. 16, 76, 109, 129, 143, 149
Johnson, S. D. 52, 136
Johnson, W. O. 17, 130
Johnston, G. F. S. 50, 136
Joiner, J. N. 4, 126
Jolliffe, P. A. 74, 142
Jones, I. 100, 148
Jones, H. E. H. 104, 148
Jones, J. 33, 133
Jones, K. M. 61, 139
Jones, R. J. 55, 56, 137
Jones, R. L. 13, 129
Jones, W. C. 38, 135
Jones, W. W. 61, 75, 138, 143

Joo, G. K. 99, 102, 147
Juillard, B. 5, 127
Julien, M. H. R. 63–65, 139
Jundulas, J. 96, 97, 147
Jung, J. 34, 134
Jurgens, S. K. 76, 143

Kabluchko, G. A. 24, 131
Kahanak, G. M. 91, 146
Kamal, A. L. 80, 144
Kanwar, J. S. 82, 145
Kapusta, G. 34, 35, 134
Karchi, Z. 89, 146
Karczmarczyk, S. J. 59, 138
Kaushik, M. P. 17, 130
Kawai, M. 57, 138
Kearns, K. R. 9, 128
Kedar, N. 89, 145
Keil, H. L. 25, 132
Kemp, T. R. 61, 138
Kennedy, E. J. 7, 127
Kennedy, P. C. 46, 135
Kenney, D. S. 21, 130
Kenney, H. 33, 133
Kessler, B. 55, 137
Kester, D. E. 4, 126
Ketring, D. L. 79, 144
Khan, A. 73, 142
Khan, A. A. 6, 127
Khan, I. A. 4, 126
Khan, R. A. 54, 137
Khosh-Khui, M. 4, 126
Khryanin, V. N. 90, 146
Khun, M. J. 79, 144
Kiang, T. Y. 4, 126
Kidd, F. 72, 142
Kihlman, B. A. 104, 148
Kilby, W. W. 9, 128
Kincaid, L. R. 46, 135
King, J. W. 79, 144
Kingston, G. 63, 64, 67, 139
Kiraly, Z. 53, 137
Kirby, B. W. 93, 97, 146
Kishi, M. 78, 144
Kissler, J. J. 78, 143
Kitamura, H. 33, 133
Kittock, D. L. 21, 53, 130, 137
Klein, I. 80, 144
Klein, M. 105, 148
Klenert, M. 37, 135
Knapp, F. W. 79, 144
Knapp, J. C. F. 91, 146
Knavel, D. E. 59, 61, 138
Knypl, J. S. 60, 138
Kochan, W. J. 80, 144
Kochman, W. 95, 97, 147
Kofranek, A. M. 47, 135
Kögl, F. 3, 126
Kojima, H. 6, 127
Komeil, A. A. 105, 148
Konnai, M. 57, 101, 138, 148
Kozlowski, T. T. 4, 55, 56, 126, 137, 138
Kozempel, M. 33, 133
Kramarovsky, E. 80, 144
Kramer, P. J. 4, 126
Krelle, E. 5, 127
Krezdorn, A. H. 28, 132
Kribben, F. J. 89, 145
Krishnamoorthy, H. N. 5, 79, 89, 127, 144–146
Kroes, R. 106, 148, 149
Krueger, W. A. 89, 146
Kubicki, B. 89, 146
Kubota, T. 24, 131
Kumar, A. 63–65, 139, 140
Kumar, S. 94, 147
Kumaran, P. M. 29, 132
Kumari, P. K. 17, 130
Kummer, V. 5, 127
Kuo, C. G. 14, 129
Kupelian, R. H. 46, 67, 135, 141
Kvale, A. 25, 132

Labanauskas, C. K. 91, 146
Lado, P. 6, 127
Lahiri, A. K. 4, 126
Laibach, F. 89, 145
Laible, C. A. 16, 129
Lakhanov, A. P. 55, 137
Lal, B. B. 25, 132
Landenauer, H. 47, 135
Lang, H. 34, 134
Langer, R. H. M. 50, 136
Lanphear, F. O. 57, 138
Larsen, F. E. 49, 136
Larson, R. A. 47, 135
LaRue, J. H. 9, 128
Laude, H. M. 50, 136
Lavee, S. 22, 26, 37, 131, 132, 135
LaVine, P. 47, 136
Law, J. 16, 129
Lawhead, C. W. 4, 126
Lawrie, I. D. 63, 67, 139
Leach, R. W. A. 64, 140
Leavitt, G. M. 78, 81, 143, 144
Leavitt, J. R. C. 99, 101, 147
Leavitt, R. A. 33, 133
Lee, C. W. 89, 146
Lee, T. H. 82, 145
Lee, T. S. 82, 145
Lee, Y. P. 64, 140
Leeper, R. W. 8, 128
Leighty, E. G. 104, 148
Leopold, A. C. 4, 50, 73, 126, 136, 142
Lessel, B. 106, 148
Levitt, G. 66, 141
Lewis, A. J. 48, 136
Lewis, L. N. 76, 91, 92, 143, 146
Lewis, T. 100, 148
Lewis, W. A. 9, 128
Li, Y. Y. 22, 46, 131, 135
Libbert, E. 5, 127
Liebermann, M. 91, 146
Lima, M. 10, 128
Limarenko, A. M. 24, 131
Limsuan, M. P. 4, 127
Linan, J. 22, 131
Lincoln, C. 11, 128
Lindoo, S. J. 91, 146
Link, M. L. 35, 134
Lipe, J. A. 74, 143
Lippert, L. F. 7, 127
Lira, E. P. 89, 90, 145
Little, C. H. A. 55, 137
Liu, F. W. 73, 142
Livne, A. 55, 137
Locascio, S. J. 61, 138
Lockwood, D. 61, 80, 139, 144
Logan, B. J. 81, 144
Lona, F. 6, 127
Loney, C. E. 80, 144
Long, C. E. 35, 134
Long, F. W. 34, 35, 134
Long, R. C. 82, 145
Looney, N. E. 29, 61, 72, 73, 75, 80, 132, 139, 142–144
Lord, W. J. 61, 72, 73, 139, 142
Loupias, S. 17, 130
Low, F. C. 70, 141
Lower, R. L. 16, 89, 129, 145
Loy, J. B. 89, 90, 146
Lucchesi, A. A. 50, 136
Luckwill, L. C. 9, 128
Lueschen, W. E. 76, 143
Lukasik, S. 30, 77, 82, 133, 143, 145
Luke, H. H. 55, 137

Lumis, G. D. 34, 47, 134, 136
Lund, S. 16, 129
Lunelli, J. 26, 132
Lusby, W. 33, 133
Lyons, J. M. 89, 90, 145

Madhava Rao, V. N. 70, 141
Madori, G. 13, 129
Mady, R. 25, 132
Maestri, N. 35, 134
Magness, J. R. 24, 131
Mahapatra, P. 4, 126
Maier, V. P. 60, 138
Mainstone, B. J. 70, 141
Maire, R. 38, 135
Malavolta, E. 59, 138
Malis Arad, S. 56, 138
Malstrom, H. L. 47, 136
Mandava, N. 33, 133
Manna, C. K. 104, 148
Mansfield, T. A. 55, 56, 137
Mansour, M. H. 52, 136
Mantel, N. 104, 148
Manzo, L. 106, 148
Marcelle, R. D. 33, 133
Maretzki, A. 65–67, 140
Marger, J. 49, 136
Mariotti, A. 61, 139
Marshall, E. R. 7, 127
Marth, P. C. 12, 25, 55, 57, 73, 78, 107, 128, 132, 137, 138, 142, 144, 149
Martin, G. C. 22, 25, 47, 61, 72, 75, 131, 132, 136, 139, 142, 143
Martin, H. 100–102, 147
Martin, W. C. 20, 47, 130, 135
Martine, J. F. 63–65, 139
Massol, R. 49, 136
Matsui, T. 33, 133
Matsukuma, I. 101, 148
Matta, F. B. 61, 138
Matthees, D. 33, 133
Mauney, J. R. 53, 137
Mavrich, E. P. 34, 137
Maw, G. A. 47, 136
Maxwell, R. C. 51, 136
Mayberry, B. D. 104, 148
Mayer, A. M. 6, 127
McArdle, R. N. 33, 133
McCarty, M. K. 95, 147
McCollum, J. P. 29, 132
McCown, M. 24, 131
McCune, S. B. 77, 141
McDavid, C. R. 13, 129
McDowell, T. 47, 135
McIntire, W. S. 95, 147
McIntosh, A. H. 54, 137
McIntyre, G. 63, 139
McIntyre, M. 47, 135
McKenzie, J. 104, 148
McKibben, G. E. 76, 143
McMeans, J. L. 47, 94, 136, 146
McMullan, E. 14, 129
McMurray, A. L. 16, 89, 129, 145
McNulty, P. J. 16, 129
McPhail, A. T. 46, 135
Meagher, M. D. 14, 129
Medan, H. 63–65, 139
Meggitt, W. F. 35, 103, 134, 148
Mehta, P. K. 78, 144
Meidner, H. 55, 137
Mellenthin, W. M. 30, 80, 133, 144
Menhenett, R. 34, 134
Menon, M. A. 70, 141
Merkle, M. E. 11, 128
Messeri, C. 4, 126
Meuzies, R. 30, 133
Meyer, M. 5, 127
Miaullis, J. B. 99, 102, 147
Micke, W. C. 25, 72, 73, 132, 142
Miele, A. 26, 132
Miller, C. H. 16, 89, 129, 145
Miller, C. O. 6, 127
Miller, C. S. 20, 105, 130
Miller, D. A. 16, 130
Miller, D. G. 16, 129
Miller, F. R. 95, 97, 147
Miller, P. M. 56, 138
Miller, S. D. 99, 100, 102, 147
Mills, H. H. 9, 61, 80, 128, 139, 144
Milo, G. E. 53, 137
Mishra, D. 56, 138
Mitchell, A. E. 24, 131
Mitchell, F. G. 81, 144
Mitchell, I. 105, 148
Mitchell, J. W. 3, 73, 107, 126, 142, 149
Mitchell, W. D. 89, 146
Mittelheuser, C. J. 55, 56, 137
Mizrahi, Y. 55, 56, 137, 138
Mochida, K. 27, 132
Mock, T. 38, 135
Modlibowska, I. 61, 139
Mohan Ram, H. Y. 89, 146
Moir, G. F. J. 70, 141
Mokashi, A. N. 4, 126
Molero, J. J. 76, 143
Monn, D. K. 82, 145
Monselise, S. P. 13, 76, 129, 143
Montalvo-Zapata, R. 63, 64, 66, 67, 140
Moore, D. C. 80, 144
Moore, J. F. 16, 130
Moore, P. H. 32, 133
Moore, R. C. 24, 25, 131, 132
Moore, R. H. 15, 17, 129
Moorhouse, J. E. 52, 136
Moraes, V. H. F. 70, 141
Mori, M. 105, 148
Morini, S. 25, 131
Morrell, G. 73, 142
Morris, J. R. 74, 142
Moshonas, M. G. 62, 139
Moss, G. E. 12, 13, 128, 129
Mukherjee, T. P. 4, 126
Muller, G. 99, 101, 147
Muller, J. C. 65, 140
Murashige, T. 73, 142
Murphy, J. J. 99, 102, 147
Murray, F. J. 106, 148
Murthy, K. N. 29, 132
Myers, H. R. 34, 35, 134

Nagarajan, M. 94, 97, 147
Nagel, C. W. 61, 139
Nair, V. K. B. 70, 141
Naito, R. 27, 132
Nakasone, L. 76, 143
Nakayama, R. M. 61, 138
Nalewaja, J. D. 99, 100, 102, 147
Namken, L. N. 34, 134
Nanda, R. 50, 136
Nandpuri, K. S. 82, 145
Narasimhan, R. 63–65, 139
Natrova, Z. 16, 130
Natti, T. A. 89, 90, 146
Naumann, K. 95, 97, 147
Nayar, N. M. 29, 132
Naylor, A. W. 15, 17, 129
Negi, S. P. 70, 141
Neidermyer, R. W. 35, 134

Nell, T. A. 34, 134
Nelson, G. S. 74, 142
Nelson, J. M. 6, 127
Nelson, J. W. 61, 139
Nelson, M. M. 25, 131
Nelson, P. M. 17, 130
Nelson, P. V. 47, 135
Nelson, W. W. 76, 143
Nemeth, G. 4, 126
Nester, P. R. 34, 134
Neto, O. G. 70, 141
Neubert, A. M. 80, 144
Ng, J. C. 9, 128
Ng, W. C. 70, 141
Nicholson, W. F. 11, 128
Nickell, L. G. 1, 4, 10, 32, 50, 63–68, 78, 126, 128, 133, 136, 139–141, 144
Nishi, Y. 105, 148
Nishijima, C. 22, 25, 47, 131, 136
Nitsch, J. P. 89, 146
Nix, L. E. 72, 142
Nolle, H. H. 35, 134
Nooden, L. D. 91, 146
Norden, A. J. 79, 144
Norman, J. C. 9, 128
Norred, W. P. 105, 148
Northrop, J. H. 104, 148
Norton, R. A. 37, 135
Nyeki, J. 30, 133
Nyenhuis, E. N. 9, 128
Nyffeler, A. 99, 101, 147

Obenauf, G. L. 81, 144
Ogata, Y. 46, 135
Ogawa, M. 33, 133
Ohl, B. 5, 127
Ohlrogge, A. J. 33, 76, 133, 143
Okatan, Y. 91, 146
Okii, M. 57, 101, 138, 148
Oku, K. 104, 148
Olsen, K. L. 72, 142
Olson, W. H. 25, 132
Olympios, C. M. 29, 132
O'Mara, R. M. 25, 132
Onitake, T. 57, 138
Oohata, J. T. 25, 132
Oota, Y. 6, 127
Opitz, K. 22, 131
Oplinger, E. S. 34, 76, 77, 133, 143
Orchard, P. W. 56, 138
Orcutt, D. M. 33, 133
Orson, P. 47, 135
Osborne, D. J. 52, 91, 136, 146
Osgood, R. V. 64, 67, 68, 140, 141
Oswald, T. H. 53, 137
Otten, G. G. 64, 140
Overdahl, C. J. 76, 143
Owens, K. W. 90, 146
Oyamada, K. 33, 133

Pakianathan, S. W. 4, 70, 126, 141
Palevitch, D. 6, 127
Pallos, F. M. 99, 100, 102, 147
Pallotta, A. J. 105, 148
Palmer, K. H. 46, 135
Pan, Y. C. 64, 140
Pao, T. P. 10, 128
Parham, M. R. 72, 142
Parmar, C. 25, 132
Parups, E. V. 34, 35, 134
Parvin, P. E. 4, 126
Patel, B. 90, 146
Patel, V. J. 90, 146
Patzold, C. 104, 148
Paul, K. B. 33, 133
Peacock, W. L. 78, 143
Peek, J. W. 99, 101, 147
Penner, D. 35, 99, 101, 103, 134, 147, 148
Penyazkova, N. V. 16, 17, 130
Peters, J. 105, 148
Peters, W. J. 72, 142
Petersen, C. E. 89–91, 145, 146
Peterson, G. E. 107, 149
Peterson, J. R. 27, 132
Peterson, L. W. 100, 101, 148
Peterson, R. H. 76, 143
Petrucelli, L. 105, 148
Petty, J. H. P. 9, 128
Pfeiffer, R. K. 67, 141
Pfrimmer, T. R. 11, 128
Pharis, R. P. 14, 129
Pharr, D. M. 89, 145
Phillips, B. A. 99, 101, 147
Phillips, D. R. 53, 137
Phillips, J. R. 11, 128
Phillips, M. R. 66, 141
Pike, L. M. 89, 90, 145
Pillai, V. B. 70, 141
Pillaiyar, P. 94, 97, 147
Pitrik, P. 77, 143
Plant, H. L. 35, 36, 134
Plaut, Z. 55, 137
Pokorny, R. 107, 149
Poling, S. M. 60, 61, 71, 138, 141
Poljakoff-Mayber, A. 6, 127
Pool, B. M. 37, 135
Pool, R. M. 29, 78, 132, 143
Poole, R. T. 38, 135
Popov, B. V. 16, 130
Porpiglia, P. J. 25, 131
Porter, C. A. 64, 65, 68, 140, 141
Porter, K. B. 16, 17, 130
Pothuluri, J. V. 80, 144
Powers, J. R. 61, 139
Prabhu, S. R. 82, 145
Pradhan, G. C. 56, 138
Prasad, D. C. 50, 136
Prakash, G. 89, 145
Preston, A. P. 49, 136
Prieto, J. 22, 131
Primot, L. 70, 141
Prince, V. E. 25, 78, 132, 144
Probst, A. H. 81, 144
Proebsting, E. L. 9, 61, 80, 128, 139, 144
Pulido, M. L. 67, 141
Puritch, G. S. 14, 129
Puttock, M. A. 35, 134
Py, C. 9, 128

Quaife, M. L. 105, 148
Qualset, C. O. 16, 17, 129, 130
Quebedeaux, B. 64, 140
Quick, W. A. 53, 137
Quinlan, J. D. 49, 136

Raafat, A. 82, 145
Raff, J. W. 78, 144
Raghavendra, A. S. 55, 56, 137
Ramanujam, T. 94, 97, 147
Ramina, A. 25, 131
Randall, D. I. 8, 128
Randhawa, G. S. 70, 141
Rane, D. A. 78, 143
Rao, I. M. 56, 138
Rao, K. C. 63, 64, 139

Rao, V. H. M. 17, 130
Rappaport, L. 7, 127
Raschke, K. 55, 56, 137, 138
Rashid. A. 4, 126
Rasi-Caldogno, F. 6, 127
Rasmussen, G. K. 22, 130
Rasmussen, S. 5, 127
Ratts, K. W. 67, 141
Rauch, F. D. 4, 126
Rawlins, T. E. 54, 55, 137
Ray, J. R. 55, 57, 137
Read, E. P. 5, 127
Reddy, C. S. 56, 138
Reed, W. 22, 131
Reeder, N. 76, 143
Rehm, S. 15, 17, 129
Reichert, B. L. 105, 148
Reid, C. P. P. 72, 142
Reimer, C. A. 24, 131
Reina, A. 22, 131
Renard, H. A. 6, 127
Renberg, L. 105, 148
Reuhanen, H. 106, 148
Rhoads, F. M. 79, 144
Rhodes, W. H. 78, 144
Rice, E. R. 69, 141
Rice, P. W. 79, 144
Richelmi, P. 106, 148
Richman, T. L. 33, 133
Richmond, A. E. 55–57, 137, 138
Richmond, P. T. 9, 128
Rieck, C. E. 35, 134
Riehl, L. A. 58, 92, 138, 146
Ries, S. K. 33, 133
Rigbey, B. 61, 139
Riker, A. J. 4, 126
Rikin, A. 57, 138
Rittig, F. R. 34, 134
Roark, B. 11, 128
Robbins, W. E. 33, 133
Roberts, D. R. 72, 142
Roberts, J. W. 24, 131
Robertson, J. 6, 127
Robinson, A. G. 51, 136
Robinson, R. W. 16, 29, 82, 89, 90, 129, 132, 145, 146
Robitaille, H. A. 24, 131
Rocha, R. F. 13, 129
Rodriguez, A. B. 3, 8, 126, 128
Rodriguez, J. G. 52, 136
Rodriguez, R. R. 34, 134
Rodriguez, S. J. 76, 143
Roethling, T. 95, 97, 147
Rogers, B. L. 25, 131
Rogers, M. N. 38, 135
Rogers, O. M. 4, 126
Rohwedder, W. K. 33, 133
Rojas-Garciduenas, M. 30, 56, 133, 138
Rom, R. C. 25, 78, 132, 144
Romani, R. J. 80, 144
Roncoroni, E. J. 25, 132
Rosario, E. L. 63–67, 140
Rosas, G. S. 16
Ross, S. D. 14, 129
Rossman, E. C. 17, 130
Rostron, H. 63, 64, 68, 139
Roth, C. B. 76, 143
Roth, J. A. 76, 143
Rousseau, G. G. 4, 126
Rowell, P. L. 16, 129
Roy, B. N. 5, 127
Roy, T. 4, 126
Rozej, B. 94, 97, 147
Rudich, J. 89, 145, 146
Rueppel, M. L. 64, 140
Rufener, J. 22, 131
Rusch, R. 20, 130
Russo, L. 73, 142
Rutledge, S. R. 82, 145
Ryan, F. J. 72, 142
Ryerson, D. K. 34, 133
Ryugo, K. 9, 61, 128, 139

Sacher, R. M. 17, 130
Sachs, R. M. 47, 135, 136
Sagaral, E. G. 33, 133
Saimbhi, M. S. 82, 145
Saini, A. D. 50, 136
Saito, T. 90, 146
Salter, P. J. 82, 145
Samish, R. M. 6, 26, 127, 132
Sams, C. E. 89, 146
Samuels, G. 63–65, 140
Sanderson, K. C. 47, 135
Sansavini, S. 49, 136
Santakumari, M. 56, 138
Santelmann, P. W. 34, 35, 79, 134, 144
Sapra, V. T. 16, 17, 129
Sarkar, A. K. 63–67, 139
Sarooshi, R. A. 22, 130
Sasseville, D. 33, 133
Satchuthananthavale, R. 70, 141
Sawyer, R. L. 7, 127
Scales, A. L. 11, 128
Schafer, D. E. 100–102, 147, 148
Scherings, H. G. 56, 138
Schiller, C. 104, 148
Schliebs, R. 82, 145
Schmalzl, K. 63, 64, 139
Schmid, A. 5, 127
Schmidt, R. R. 82, 145
Schneider, G. 35, 134
Schneider, G. W. 24, 38, 131
Schott, P. E. 34, 35, 134
Schreader, W. R. 25, 132
Schroeber, M. 34, 134
Schuitemaker, F. 63, 64, 140
Schwarz, O. J. 72, 142
Schwetz, B. A. 106, 148
Scora, R. W. 91, 146
Scott, K. R. 78, 144
Seeschaaf, K. W. 70, 141
Seeyave, J. 9, 128
Seidel, M. C. 17, 130
Sekhar, B. C. 70, 141
Selevicius, A. 96, 97, 147
Sell, H. M. 90, 146
Selman, I. W. 53, 137
Seltmann, H. 46, 135
Sethuraj, M. R. 70, 141
Sethuraman, S. 94, 97, 147
Shafir, N. 76, 143
Shahin, H. 13, 129
Shanmugavelu, K. G. 70, 141
Shannon, S. 29, 89, 132, 145
Sharma, G. C. 16, 17, 129
Sharma, J. K. 17, 130
Sharma, R. A. 63–65, 139
Sharma, R. C. 7, 127
Sharma, R. K. 63–65, 139
Sharma, S. R. 63–65, 139
Sharples, G. C. 6, 127
Sharples, R. O. 61, 139
Shaw, P. E. 62, 139
Shaw, W. C. 94, 97, 147
Shawa, A. Y. 61, 139
Sheets, W. A. 37, 135
Shigeura, G. 76, 143
Shimkin, M. B. 105, 148
Shimshi, D. 55, 56, 137
Shively, E. A. 61, 139
Shmueli, E. 55, 137
Shu, L. J. 47, 135
Shukla, S. N. 59, 138
Shulman, Y. 37, 135
Sibbett, G. S. 22, 25, 47, 131, 132, 136
Siddiqui, B. A. 4, 126
Siemer, S. R. 78, 144

Sill, L. Z. 47, 135
Siller, A. 30, 133
Sim, G. A. 46, 135
Sims, E. T. 78, 144
Sims, W. L. 82, 89, 90, 145
Simmons, C. S. 14, 129
Singh, B. 63, 65, 139
Singh, G. 73, 94, 142, 147
Singh, K. 63, 65, 139
Singh, S. P. 4, 17, 126, 130
Singh, T. P. 16, 130
Singh, U. R. 73, 142
Singletary, G. W. 33, 133
Singmaster, J. A. 9, 128
Sinha, A. K. 53, 137
Sitton, B. G. 9, 128
Slatyer, R. O. 55, 56, 137
Slife, F. W. 99, 102, 147
Smagula, J. M. 61, 139
Smith, G. A. 16, 129
Smith, O. 7, 127
Smith, R. J. 94, 97, 98, 106, 147, 148
Smith, T. S. 61, 138
Smock, R. M. 83, 145
Smrz, J. 77, 143
Sochilin, E. C. 16, 17, 130
Soethara, A. 17, 130
Soltesz, M. 30, 133
Son, E. R. 82, 145
Soopramanien, G. C. 63–65, 139
Sorvari, T. E. 106, 148
Sotnik, V. M. 16, 129
Southorn, W. A. 70, 141
Spaulding, D. W. 46, 135
Spencer, G. F. 33, 133
Splittstoesser, W. E. 82, 89, 145, 146
Sponagle, A. G. 82, 145
Srinivasan, C. 73, 142
Srivastava, B. S. 53, 137
Srivastava, S. C. 63, 65, 139
Stackhouse, S. S. 74, 143
Stacy, E. M. 104, 148
Stahly, E. A. 30, 73, 133, 142
Stang, E. J. 30, 73, 133, 142
Stark, R. 74, 143
Starke, G. R. 16, 46, 57, 129, 135
Stebbins, R. L. 24, 131
Steffens, G. L. 33, 46, 47, 82, 133, 135, 139
Stelzer, L. R. 93, 97, 146
Stembridge, G. E. 9, 25, 73, 78, 128, 131, 132, 142, 144
Stephenson, G. R. 99, 101, 102, 147
Stergios, B. G. 74, 143
Stewart, W. S. 77, 143
Stoddard, E. M. 56, 138
Stoddard, J. L. 91, 146
Stoller, E. W. 94, 97, 98, 147
Stores, E. E. 105, 148
Stoskopf, N. C. 16, 18, 129
Strydom, D. K. 61, 139
Stuart, N. W. 12, 47, 128, 135
Stupiello, J. P. 50, 136
Sturelid, S. 104, 148
Stutte, C. A. 34, 81, 82, 134, 145
Sud, A. 50, 136
Suge, H. 90, 146
Sullivan, D. T. 9, 128
Sullivan, T. D. 35, 65, 134
Sumiki, Y. 32, 133
Supriewska, J. H. 82, 145
Suquet, J. 49, 136
Suryanarayana, T. K. 82, 145
Suttner, D. L. 34–36, 134
Suzuki, K. 25, 132
Swaminanthan, M. S. 17, 130
Swamy, P. M. 56, 138
Sweeley, C. C. 33, 133
Sweet, C. P. M. 63, 64, 68, 139
Sympson, R. L. 4, 126
Szabo, Z. 25, 132
Szirmai, J. 53, 137
Szkrybalo, W. 47, 135

Tabbak, C. 90, 146
Tafazoli, E. 4, 29, 126, 132
Tahori, A. S. 52, 136
Takahashi, D. T. 50, 64–68, 136, 140, 141
Takahashi, H. 90, 146
Takahashi, M. T. 104, 148
Takematsu, T. 57, 101, 138, 148
Tanaka, I. 4, 127
Tanimoto, T. T. 10, 32, 63–67, 128, 133, 139–141
Tapio, D. D. 82, 145
Tarasenko, M. T. 4, 126
Tasaki, M. 78, 144
Tayler, R. S. 70
Taylor, J. B. 22, 131
Terent'ev, A. G. 17, 130
Teshima, A. 64, 67, 140
Tewari, M. N. 59, 138
Thakur, A. K. 63–67, 139
Thakur, D. R. 25, 132
Theiss, J. C. 105, 148
Theurer, J. C. 16, 129
Thimann, K. V. 3, 4, 6, 91, 126, 127, 146
Thomas, G. J. 35, 64, 134, 140
Thomas, H. 91, 146
Thomas, J. K. 4, 126
Thomas, R. O. 11, 53, 128, 137
Thomas, T. H. 6, 82, 127, 145
Thomas, V. M. 99, 102, 147
Thompson, A. H. 24, 25, 131
Thompson, J. B. 61, 139
Thompson, M. J. 33, 133
Tianco, A. P. 63, 64, 67, 140
Timm, H. 7, 127
Timmons, P. R. 105, 148
Tisza, A. 30, 133
Tjia, B. 48, 136
Tobitsuka, J. 33, 133
Tolbert, N. E. 6, 127
Tolla, G. E. 89–91, 145, 146
Tombesi, A. 22, 131
Tompsett, P. B. 14, 129
Torabi, M. 78, 144
Toth, M. 25, 132
Troncoso, A. 22, 131
Trupp, C. R. 16, 129
Tsai, S. D. 54, 137
Tso, T. C. 46, 135
Tucker, D. J. 47, 136
Tukey, H. B. 107, 149
Tukey, L. D. 29, 132
Tullis, E. C. 94, 97, 98, 147
Tupy, J. 70, 141
Turgeon, A. J. 35, 134
Tweedy, J. A. 34–36, 134
Tyler, R. H. 72, 73, 142

Ulland, B. M. 105, 148
Unrath, C. R. 24, 73, 131, 142

Vaadia, Y. 55, 137
Vail, P. V. 53, 137

Valerio, M. G. 105, 148
Valio, I. F. M. 13, 129
VanBennekom, J. L. 17, 130
VanBlaricom, C. O. 73, 142
VanDam, R. 17, 130
Vandemark, J. S. 82, 145
Van der Meer, O. P. 17, 130
Vanderventer, J. W. 34–36, 134
VanEmden, H. F. 51, 55, 136
van Esch, G. J. 106, 149
van Overbeek, J. 29, 132
VanSchaik, P. H. 81, 144
VanSteveninck, R. F. M. 55, 56, 137
Varenitsa, E. T. 16, 129, 130
Vega, J. 37, 134
Vega, N. 65, 66, 140, 141
Veinbrants, N. 24, 131
Velez, A. 63–65, 140
Vereecke, M. 48, 136
Verner, L. 80, 144
Verschuuren, H. G. 106, 148, 149
Vienravee, K. 38, 135
Villareal, R. L. 6, 57, 127, 138
Villemur, P. 49, 136
Vines, H. M. 61, 80, 139, 144
Vitagliano, C. 22, 25, 131
Vlitos, A. J. 63, 67, 139
Vogt, H. E. 16, 17, 129, 130
von Kuhn, H. 82, 145
von Stillfried, H. 20, 130

Waddle, B. A. 11, 128
Waggoner, P. E. 55, 56, 137
Wai, S. T. 70, 141
Waisel, Y. 55, 56, 138
Waister, P. D. 37, 135
Walhood, V. T. 53, 94, 137, 146
Walker, B. 63–65, 140
Walker, E. K. 82, 145
Wall, M. E. 46, 135
Walther, H. R. 47, 135, 136
Wang, C. Y. 30, 80, 91, 133, 144, 146
Wang, R. C. 16, 129
Wani, M. C. 46, 135
Warner, H. L. 16, 17, 129, 130
Warthen, J. D. 33, 133
Watschke, T. L. 34, 35, 134
Weakley, M. L. 66, 140
Weaver, R. J. 4, 7, 9, 26, 29, 37, 77, 78, 126–128, 132, 135, 143
Wee, Y. C. 9, 128
Weerasinghe, T. C. 70, 141
Wehner, D. J. 34, 35, 134
Weinbaum, S. A. 25, 131
Weinberger, H. H. 78, 144
Weiser, M. 76, 143
Welbank, P. J. 34, 133
Wells, W. H. 66, 141
Went, F. W. 3, 126
Wert, V. F. 33, 133
Weston, G. D. 82, 145
Westwood, M. N. 24, 131
Weybrew, J. A. 82, 145
Wheeler, J. E. 6, 127
Whenham, R. 53, 137
Whigham, D. K. 94, 97, 98, 147
Whisler, J. 22, 131
Whitacker, T. W. 16, 129
Whitty, E. B. 46, 135
Wickenden, M. F. 61, 139
Widmoyer, F. B. 9, 128
Wiese, A. F. 16, 130
Wilcox, G. E. 82, 145
Wilcox, M. 22, 46, 131, 135
Wilczynski, H. 82, 145
Wilfret, G. J. 13, 34, 48, 129, 134, 136
Wilkes, L. H. 52, 136
Wilkinson, R. E. 25, 131
Will, H. 35, 134
Williams, A. H. 94, 147
Williams, F. J. 94, 97, 98, 147
Williams, M. W. 22, 24, 25, 30, 61, 72, 73, 75, 131, 133, 139, 142, 143
Wilson, R. G. 103, 148
Wilson, W. C. 21, 22, 130, 131
Witts, K. J. 34, 133
Wittwer, S. H. 6, 7, 17, 33, 79, 89, 127, 130, 133, 145, 146
Wolter, K. E. 72, 142
Woltz, W. G. 82, 145
Wood, R. A. 63, 64, 67, 68, 139
Wood, R. K. S. 53, 137
Woodbury, W. 33, 133
Woolhouse, H. W. 91, 146
Worku, W. 61, 138
Worku, Z. 80, 144
Worley, J. F. 33, 46, 133, 135
Worley, R. E. 47, 136
Wright, S. T. C. 55, 137
Wu, C. H. 34, 35, 79, 134, 144
Wu, W. L. 9, 128
Wunsche, V. 57, 138
Wycherley, P. R. 70, 141
Wyllie, T. D. 53, 137
Wynne, J. C. 79, 144

Xiloyannis, C. 25, 131

Yabuta, T. 32, 133
Yaeger, J. F. 72, 73, 142
Yamaguchi, M. 4, 77, 126
Yamakawa, R. M. 4, 126
Yamamura, H. 27, 132
Yang, P. C. 10, 63–67, 128, 140
Yates, R. A. 63–65, 140
Yeager, J. T. 25, 132
Yen, S. T. 6, 127
Ylatalo, M. 4, 126
Ylitalo, P. 105, 148
Yokoyama, H. 60, 61, 71, 138, 141
Young, E. 25, 132
Youngner, V. B. 34, 35, 134
Yrarrazaval, F. 9, 128
Yu, K. S. 24, 131
Yu, P. K. 46, 135

Zak, C. D. 89, 90, 146
Zaki, M. E.-S. 82, 145
Zamora, O. B. 63–67, 140
Zebrowski, Z. 57, 138
Zeidler, G. 52, 136
Zeisig, H. C. 99, 100, 147
Zelitch, I. 55, 56, 137, 138
Zevite-Kulvetiene, Z. 54, 137
Zieslin, N. 13, 129
Zigas, R. P. 6, 127
Zimmerman, P. W. 3, 107, 126, 149
Zimmerman, R. H. 47, 135
Zocca, A. 49, 136
Zook, F. 35, 134
Zschoche, W. C. 64, 140
Zucconi, F. 25, 131
Zucconi, R. 22, 131
Zukel, J. W. 35, 36, 134
Zuluaga, E. M. 26, 132

Subject Index

abortion 15
abscisic acid 4, 6, 19, 55–57, 73, 79, 91
abscission 19, 21, 22, 25, 27, 28, 61, 73, 75, 83, 91, 94, 112, 113, 116, 117, 119
abscission zone 19
absorption, herbicide 103
acaricide 23, 113
Acaron 113
Accelerate 98
acephate 11, 115
acetophenones 46
acetylene 3, 8, 110
AGP-322 121
ACR-1308 35, 64
Acti-Aid 116
Actidione 65
adenoma bioassay 105
ADI 106
adjuvants 20
Agrotect 114
Agroxone 113, 119
AIB 64, 110
alachlor 56, 99, 101, 111
Alanap 121
Alar 123
alcohol, fatty 48
alcohol, long-chained 33
alfalfa 16, 33
aliphatic alcohols, methyl esters 47
alkarylpolyoxyethylene glycol 48
alkenylsuccinic acid 56, 110
alkylene diamine 57
alkylthiocarbamates 100
l-allyl-1(3,7-dimethyloctyl) piperidinium bromide 34
almond 9
Alsol 22, 112
Amaize 111, 117
Amchem 66–329 64
American bollworm 52
Ames test 104
Ametrex 98
ametryn 94, 98, 117
Amid-Thin 23, 120
p-aminobenzenesulfonyl urea 64, 110
4-amino-(1,1-dimethylethyl)-3-(methylthio)-1,2,4-triazin-5(4)-one 101
aminoethoxyvinylglycine (AVG) 30, 80, 90, 110
2-amino-4-methoxy-3-butenoic acid 36, 110
2-amino-6-methylbenzoic acid 30, 35, 57, 64, 82, 110
aminomethylphosphonic acid 64, 110
6-aminopenicillanic acid 64, 110
2-aminopurine 104
6-amino-o-toluic acid 64
3-amino-1,2,4-triazole 110
4-amino-3,5,6-trichloropicolinic acid 64, 110
amitrole 52, 110
ammonium ethyl carbamoyl phosphonate 64
ammonium isobutyrate 64, 78, 110
Amo-1618 57
AMPA 64, 110
amphibians 104
amylase 32, 125
ancymidol 34, 36, 38, 82, 114
anisomycin 66, 119
Anna-Amide 23
Anofex 114
anthocyanin 75
antibiotic 110, 119, 121
antibiotic, antifungal 21, 116
antidotes, herbicide 99–101, 113–115, 119, 124
anti-feedant, insect 110, 121
antifungal antibiotic 21, 116
anti-ripening agents 83, 115
anti-transpirant 55, 56, 110–113, 117–119, 121–123
antitumor agent 46
6-APA 64, 110
aphid 51, 52
apical dominance 45, 47, 48, 79
apple 4, 9, 22–25, 30, 31, 49, 52, 61, 72, 83, 108
apricot 73
Aquathol 98
arborvitae 4

A-Rest 36, 114
Arkotine 114
aroids 13
arrowing 10
arsenic acid 20, 94, 97, 105, 110
arthritis drug 119
ash 56
asulam 66, 120
Asulox 66, 120
Atlacide 97
atrazine 99, 101
Atrinal 35, 120
auxin 1, 4, 6, 8, 9, 19, 28, 38, 50, 80, 83, 89–91
AVG 110
axil 50
axillary buds 45, 115
Azalea 47, 48
6-azauracil 67, 123
azinphosmethyl 116

B-9 123
bacitracin 67, 110
bactericide 111
bacteriophage 104
banana 28, 73
banana pepper 80
BAP 50, 111
barban 99, 101
bark, tree 70
barley 16, 32, 42, 50, 56, 57
barnyard grass 18
BAS-0830 116
Basanite 111, 117
Bay-5072 116
Bay-17147 116
Bay-22555 116
bean 32, 39, 56, 93, 96
–, blight, southern 15
–, broad 51, 52
–, common 5
–, lima 57
–, mung 5
–, navy 57
–, pole 57
–, runner 5
–, snap 52, 57
–, wax 52
–, winged 4
begonia 47
beet 9
Beet-Kleen 112
bell pepper 61
benomyl 54
benzyl hydrazone 101
N-(2′)-benzimidazolyl-1,8-naphthalimide 102
N-benzoyl-N-(3,4-dichlorophenyl)-aminopropionic acid 64, 111
6-benzyladenine 30, 31, 61, 73, 91, 111
6-benzylaminopurine (BAP) 50, 90, 111
6-benzylamino-9-(tetrahydropyran-2-yl)-9H-purine 50, 110
benzylhydrazones 99
ber 90
Bermat 113
berries, black 74
–, blue 74
–, cane 74
–, coffee 76
–, grape 26, 27, 29, 78
–, rasp- 37, 57, 74
biennial bearing 25
bindweed 107
biocide 118
biomass 84
bioregulation 2, 60
bioregulator 2, 60
bipyridylium quaternary ammonium compounds 106
birch 56
N,N-bis(phosphonomethyl)glycine 64
bis(N,O-trifluoroacetyl)-N-phosphonomethyl glycine 64, 124
Biuret 66, 118
blackberry 74
blight, bean, southern 15
blueberry 74
blueberry, highbush 74
–, rabbiteye 74
B-Nine 123
BNOA 121
BOH 118
boll, cotton 10, 11, 13, 31, 37, 44, 93, 94, 111, 113, 114, 116
–, tree 72
– weevil 53
bollworm, American 52
–, pink, cotton 52, 53
borer, sugarcane 51
Bougainvillea 4
box elder 57
Brassica rapa 6
brassinolide 33, 123
broad bean 51, 52
broccoli 91
bromacil 64, 111
bromeliads 8
5-bromo-3-sec-butyl-6-methyluracil 64, 111
5-bromouracil 104
Brush Killer 124
Brussels sprouts 17, 48, 51

BTS 34-273 66, 121
Bualta 67, 122
buds 7, 9, 12, 37, 45, 48, 70, 112, 116, 123
budworm, tobacco 53
bulbs 7
bullhead 13
butachlor 101
2-t-butylamino-4-ethylamino-6-methylthio-s-triazine 101
2-sec-butyl-4,6-dinitrophenol 98, 111

C-8514 113
C-19490 36
cabbage 9, 17, 33, 57
cabbage aphid 51
–, Chinese 6, 57
cacodylic acid 20, 65, 104, 116, 118
calcium 59
– polysulfide 22
calix 30
Callistemon 47
Camellia 12
camphene, chlorinated 111
Camphoclor 111
Camptotheca acuminata 46
camptothecin 46
Canada thistle 103
cane 37
– berries 74
Cannabis sativa 89
cantaloupe 5
carbamyl urea 66, 118
carbaryl 23, 24, 121
carbofuran 77
carboxin 99, 115
carcinogen 104–106
carnation 9, 57, 91
carob 4
carotene 61
carotenoid 60, 61, 71, 76
carrot 9, 82
Carvadan 52
cashew nut 29
cauliflower 17, 82
CCC 65, 112
celery 6
cell 19, 34, 37, 72, 118
cellulase 19
CEPA 112
Cepha 23, 64, 112
cercosporiosis 54
cereals 16, 17, 33, 50, 101
Cetrimide 65, 118
cetyltrimethylammonium bromide 65, 118
CF-125 119
CGA-13586 112
CGA-15281 23, 24, 111
CGA-17020 35
CGA-24705 35, 111
charcoal rot 53
CHE-8728 113
Chemsect 23, 117
Chemox 117
cherry 25, 28, 49, 73, 75, 86
–, sour 75
–, sweet 25, 49, 61, 75
chili, pepper 61
Chinese cabbage 6, 57
chlordimeform 11, 113
chlorflurenol 29, 34, 35, 50, 53, 56, 82, 103, 119
chlormequat 6, 10, 13, 29, 30, 33, 37, 43, 51–57, 59–61, 65, 68, 69, 76, 79, 82, 89, 90, 92, 105, 112
chloroacetanilides 99, 100
chloroacetic acid, sodium salt 94, 98
2-chlorobenzoic acid 64, 111
1-(2-chlorobenzyl)-3-carboxy-4,6-dimethylpyrid-2-one 16, 111
4-chloro-2-butynyl-m-chlorocarbanilate 101
2-chloro-3-(3-chloro-2-methylphenyl)propionitrile 79
2-chloro-2′,6′-diethyl-N-(butoxymethyl)acetanilide 101
2-chloro-2′,6′-diethyl-N-(methoxymethyl)acetanilide 101, 111
2-chloroethanethiophosphonic acid dichloride 82, 111
2-chloro-4-(ethylamino)-6-(1-cyano-1-methylethylamino)-1,3,5-triazine 101
2-chloroethylaminodi(methylphosphonic acid) 64, 111
2-chloro-4-(ethylamino)-6-(isopropylamino)-s-triazine 101
β-chloroethylmethyl-bis-benzyloxy silane 23–25, 111
2-chloro-N-(2-ethyl-6-methylphenyl)-N-(2-methoxy-l-methylethyl)acetamide 35, 101, 111
2-chloroethylphosphonic acid 16, 23, 64, 97, 112
2-chloroethyltrimethylammonium chloride 65, 112
2-chloroethyl-tris-(2-methoxyethoxy)silane 22, 112
chloro-IPC 112
3-chloroisopropyl-N-phenylcarbamate 25, 112
2-chloromercuri-4,6-dinitrophenol 56, 113
5-chloro-3-methyl-4-nitro-1H-pyrazole 21, 27, 113

4-chloro-2-methylphenoxyacetic acid 113
4-chlorophenoxyacetic acid 28, 105, 113
α-(p-chlorophenoxy)isobutyric acid 80, 113
3-chlorophenoxy-α-propionamide 25, 113
2-(3-chlorophenoxy)propionic acid 88
1-(p-chlorophenyl)-1,2-dihydro-4,6-dimethyl-2-oxo-nicotinic acid, sodium salt 16, 82, 113
3-(p-chlorophenyl)-1,1-dimethyl urea 113
3-(p-chlorophenyl)-6-methoxy-s-triazine-2,4-(1H,3H)-dione, triethanolamine salt 16, 113
2-(4-chlorophenylthio)triethylamine hydrochloride (CPTA) 60, 113
chlorophyll 61, 75, 76, 114
2-chloro-4-quinoline carboxylic acid 16, 57, 113
5-chloro-4-quinoline carboxylic acid 57
chlorothalonil 22
5-chloro-2-thenyl-tri-n-butylphosphonium chloride 65, 113
N′-(4-chloro-o-tolyl)-N-,N-dimethylformamidine 113
chlorphonium chloride 34, 52, 123
chlorpropham 112
chromosome 104
chrysanthemum 34, 47, 48, 91
CIPC 112
citrus 6, 12, 19, 21, 27, 28, 56, 60–62, 71, 75, 91
coccinellid larvae 51
coffee 76
– berries 76
– trees 27
color 60, 71–76, 78, 80–82, 92, 94, 112
copper sulfate 22
Cordyline terminalis 14
coriander 89
corn 15–17, 32, 56, 57, 76, 77, 100–102, 111, 117
Cotoneaster 47
cotton 6, 10, 11, 13, 16, 19, 20, 26, 31, 34, 37, 43, 44, 52–57, 93, 97, 100, 105, 108, 111, 113, 115
– boll 10, 11, 13, 19, 31, 93, 111, 113, 114, 116
– leafworm 52
cotyledon 60
CP-41845 64
4-CPA 104, 113
4-CPP 104
CPTA 60, 61, 113
crabgrass 18
cranberry 61
creasing 76
Credazine 66, 120
Cristoxo 111
crown 30, 88
Cryptomeria 4
60-CS-16 120
CTAB 65, 118
cucumber 6, 29, 56, 57, 77, 89, 90
–, pickling 29
cucurbits 6, 15–17, 77, 89, 90
Cupressaceae 14
Cuprimine 66, 119
curd, cauliflower 82
curing, tobacco 62, 82
cuttings 4, 5, 70
cutworm, greasy 52
cyanazine 101
α-(cyanomethoxamino)benzacetonitrile 99, 101, 113
3-cyclohexene-1-carboxylic acid 65, 113
cycloheximide 21, 22, 65, 80, 91, 116
α-cyclopropyl-α-(p-methoxyphenyl)-5-pyrimidine methanol 36, 114
Cycocel 65, 112
Cydonia 4
cytokinin 1, 6, 30, 31, 50, 53, 55, 91
cytoplasm 15

2,4-D 4, 8, 29, 51–53, 59, 61, 70, 73, 80, 83, 91, 99, 103–105, 107, 114
2,4-D amine 59
daffodil 91
daminozide 6, 9, 12, 13, 24, 25, 29, 30, 34, 37, 52, 54–57, 59, 72–76, 78–82, 86, 91, 123
Datura arborea 56
DDT 11, 114
decanoic acid, methyl ester 47
De-Cut 115
Dedelo 114
DEF 94, 97, 124
De-Fol-Ate 98
defoliant 19–21, 26, 52, 93, 105, 116–119, 121, 122, 124
defoliation 10, 20, 21, 53, 93
De-Green 124
desert locust 52
Des-I-Cate 98
desiccant 19, 20, 52, 93–98, 110–112, 114–119, 121, 122, 124
Desiccant L-10 97, 110
desiccation 20, 53, 93–98
determinant, plant type 81
Dexon 99, 101, 116
Dextrone 115
dialkylaminoethylamides 83
diallate 101
N,N-diallyl-2,2-dichloroacetamide 99, 101, 102, 114

N,N-diallyl-
–, 2-dichloroacetanilide 99, 114
diapause 52, 53
diazosulfonates 99
1-dibromoacetyl-2,5-dimethylpyrrolidine 102
dicamba 103
dicarbam 23
dichlorflurenol-methyl 35, 120
5-(2,3-dichloroallyl)-diisopropylthiocarbamate 101
2′,4′-dichloro-1-cyanoethanesulfonanilide 61, 114
2,4-dichloro-5-fluorophenoxyacetic acid 16
dichlorodiphenyltrichloroethane 114
2,4-dichloro-5-fluorophenoxyacetic acid 114
2,3-dichloroisothiazole-5-carboxylic acid 53, 114
2,3-dichloro-6-methylbenzoic acid 65, 114
2,3-dichloro-2-methylpropionic acid, sodium salt 16, 114
2,5-dichloro-4-nitrophenol 95, 97, 114
2,4-dichlorophenoxyacetic acid 4, 8, 114
2-(3,4-dichlorophenoxy)triethylamine 71, 114
N-(3,4-dichlorophenylcarbamoyl)-N-methylglycine monohydrate 101
3-(3,4-dichlorophenyl)-1,1-dimethyl urea 115
N,N-diethylnonylamine 61, 115
2,3-dihydro-5,6-dimethyl-1,4-dithiin-1,1,4,4-tetraoxide 95, 115
2,3-dihydro-5,6-diphenyl-1,4-oxathiin 35
6,7-dihydropyrido(1,2-a:2′,1′-c)-pyrazidinium dibromide 97, 115
dihydroimidazoisoindolediones 46
5,6-dihydro-2-methyl-1,4-oxathiin-3-carboxanilide 99, 102, 115
2,3-dihydro-5-(4-methylphenyl)-6-phenyl-1,4-oxathiin-4-oxide 35, 115
1,2-dihydro-3,6-pyridazine-dione 17, 35, 115
N-(2,3-dihydroxy-1-propyl)-N-phosphonomethyl glycine, disodium salt 65, 115
diisobutylphenoxyethoxyethyldimethylbenzyl-ammonium chloride 65, 115
dikegulac sodium 34, 35, 47, 74, 120
dill 9
dimequat 104
O,S-dimethylacetylphosphoramidothioate 115
2-(β-dimethylaminoethoxy)-4-(3′,4′-dichlorophenyl)thiazole hydrochloride 65, 116
dimethylaminomaleamic acid 12, 116
[4-(dimethylamino)phenyl]-diazenesulfonic acid, sodium salt 116
dimethylarsinic acid 65, 116
1,1′-dimethyl-4,4′-bipyridiniumdichloride 97, 101, 116
2-1-[2,5-dimethylphenyl]-ethylsulfonyl-pyridine-N-oxide 36, 116
N,N-dimethylglycine 65, 116
4,4-dimethylmorpholinium chloride 10, 53, 54, 116
O,O-dimethyl-S-[(4-oxo-1,2,3-benzotriazin-3(4H)-yl)-methyl]phosphorodithioate 116
3-(2-[3,5-dimethyl-2-oxocyclohexyl]-2-hydroxyethyl)glutarimide 65, 116
1,1-dimethylpiperidinium chloride 34, 116
N-2,4-dimethyl-5-(trifluoromethyl)sulfonylaminophenyl acetamide 35, 65, 116
Dinitro 111, 117
4,6-dinitro-sec-butylphenol 37, 98, 117
4,6-dinitro-o-cresol, sodium salt 23, 24, 117
dinitrocyclohexophenol 24, 117
2,4-dinitrophenol 56, 117
dinoseb 37, 76, 77, 94, 95, 98, 104, 111, 117
diphenyl ethers 46
diquat 10, 94–97, 115
disease 15, 47, 51, 53, 55, 114, 116, 118, 120, 121, 123, 125
disugran 66, 69, 120
dithiocarbamates 99
diuron 10, 104, 115
DNBP 111, 117
DNC 23
DNOC 23, 24, 104, 117
DNP 117
dormancy 6, 7, 118
Douglas fir 14
Dowicide 121
Dowicide G-ST 98
Dow Sodium TCA 98
DPX-1840 50
DPX-3778 16, 113
Dracaena spp. 14
drought avoidance 55
– resistance 55, 56, 79
– tolerance 55
Dynex 115

Easy-Off-D 124
ecdysis 52
eggplant 4, 28, 29
EL-531 36, 114
elder, box 57
elemicin 62
Elgetol 23, 117
Embark 35, 65, 116

embryo 7, 90
endothall 20, 21, 66, 94, 98, 104, 121
environmental pollution 69
- safety 104
- stress 55, 108
enzyme 3, 4, 32, 70, 71, 79, 105
enzyme, proteolytic 70
EPTC 99, 102
essential oil 61
ethanedialdioxime 22, 117
ethanolamine-p-nitrobenzenesulfonyl urea 36, 117
ethene 117
ethephon 8, 16, 18, 21–25, 27, 29–31, 34, 37, 42, 50, 61, 62, 64, 68–70, 72–76, 78–82, 84–87, 89, 90, 95, 97, 103, 105, 112
ethers, diphenyl 46
–, phenolic 61, 62
4-ethoxy-1-(p-tolyl)-s-triazine-2,6(1H:3H)dione 33, 117
Ethrel 16, 23, 64, 97, 112
2-(ethylamino)-4-isopropylamino-6-methylthio-s-triazine 117
ethyl-5-chloro-1H-indazol-3-acetic acid methyl ester 25, 117
ethyl-5(4-chlorophenol)2H-tetrazole-2-yl acetate 47, 49, 79, 118
5-ethyldipropyldithiocarbamate 102
ethylene 1–3, 8, 19, 24, 61, 68, 70, 80, 90, 117
S-ethylhexahydro-1H-azepine-1-carbothioate 102
eugenol 62
Euonymus 47
Euphorbia 71
Evik 71
E-Z-Off 98

fatty acid, methyl esters 46–49
fatty alcohols 48
fenaminosulf 116
ferrous sulfate 101
Festuca rubra 56
fiber, cotton 19, 34
Ficus 4
fig 28
Flagecidin 66, 119
flavoring agent 118
Florel 112
flower 9–13, 15, 21, 25, 26, 30, 34, 49, 81, 89–91, 112, 116, 123
flowering 3, 8–14, 25, 27, 29, 32, 33, 77, 79, 89, 90, 110, 112–115, 117, 118, 123, 125
fluoridamid 34, 35, 67, 124
N-(m-fluorobenzyl)phthalimide 17, 118
4-fluoro-2,6-dichlorophenoxyacetic acid 17, 118
flurenol 35, 120
Folex 124
foolish seedling 32
forest trees 4, 14
Fos-Fall-A 124
frost injury 57
– resistance 57
fruit 9–12, 19, 21, 22, 25, 27–30, 32, 37, 49, 60–62, 70–76, 78–83, 90–92, 112–114, 116–124
– development 28, 29
Fruit Fix 23
– set 12, 13, 23, 27–30, 78, 110, 112–114, 118–121, 123, 125
– shape 30, 31, 90, 111
– size 28, 73, 75, 78, 81, 88, 92
– stone 9, 25, 28, 73
Fruitone 23
Fruitone-A 124
Fruitone-N 120
Fruitone-T 124
Fuchsia 47
Fundal 113
fungicide 1, 23, 53, 111, 116, 117, 119
fungus, late blight 95
6-furfurylaminopurine 118
Fusarium wilt 53
fusicoccin 6
FW-450 16, 114

G-34162 117
GA_3 17, 40, 125
GA_4 14, 57, 73, 90
GA_7 14, 57, 73, 90
gametocide 15–18, 111–115, 118, 119, 122, 125
gardenia 4
Garfene 116
garlic, wild 103
geranium 57
germination 6, 57, 77, 112, 118, 123, 125
gherkins 89
Gibberella fujikuroa 32
gibberellic acid, gibberellins 1, 6, 7, 9, 12–14, 17, 25–33, 37–42, 50–54, 57–59, 61, 69, 73–81, 87, 89–92, 125
gibberellin mixtures 14, 57, 73, 90
Gibrel 125
Gib-Tabs 125
girdling 29
glyoxime 22, 117
glyphosate 36, 67–69, 94, 95, 97, 122
glyphosine 35, 64, 68, 69, 77, 122
gourd 89

grain 29, 30, 32, 33, 74, 94
–, sorghum 94, 95, 97, 99–101
Gramoxone 97, 116
grapes 4, 5, 9, 17, 26–29, 32, 37, 41, 57, 61, 77, 78, 125
–, seedless 78, 125
grapefruit 61, 87, 91
grass 35, 36, 38, 50, 99, 107
–, barnyard 18
–, crab 18
–, sod-former 50
–, tufted 50
–, turf 34
–, zoysia 42
greasy cutworm 52
Gro-Sol 125
guar 96, 97
guayule 71
Guthion 11, 116

H-722 66, 120
haloethanephosphonic acids 8
harvest, harvesting 1, 10, 19, 22, 63, 71, 72, 74, 75, 78–81, 91, 93, 96
– aids 19, 95, 96
–, stripper 93
hemp 90
herbicide 1, 23, 51, 60, 68, 95, 99–103, 105, 107, 110–122, 124
– absorption 103
– antidotes 99–102, 113–116, 119, 121, 124
– damage 99
– injury 99
– safeners 99–102
Hevea brasiliensis 69–71, 85
hexadecyltrimethylammonium bromide 65, 118
Hibiscus 4
highbush blueberry 74
Hill reaction 99
Hi-Yield Desiccant 97
holly 12, 47
Hormodin 118
human safety 104
Hyamine-1622 65, 115
hybrid vigor 15
hydrazine 104
Hydrothol 98
hydroxydimethylarsine oxide 118
1-hydroxy-1,1-ethane diphosphonic acid 65, 118
β-hydroxyethylhydrazine 8, 118
4-hydroxy-3-methoxy benzaldehyde 66, 118
8-hydroxyquinoline 56, 118
1-hydroxytriacontane 118
Hyvar-X 64, 111

IAA 53, 118
IBA 4, 5, 29, 82, 118
imidodicarbonic diamide 66, 118
Impatiens balsamina 6
indeterminant plant type 81
indoleacetic acid (IAA) 3, 30, 56, 73, 80, 106, 118
indolebutyric acid (IBA) 4, 5, 118
inhibitor 1, 2, 7, 9, 20, 29, 34, 37, 50, 80
insect 51, 52, 112, 114–116, 119, 121, 123, 125
insecticide 1, 11, 23, 24, 111, 113–117, 119–121, 123, 125
iodoacetic acid 21, 119
irrigation 10, 79
isoaureomycin 66, 119
isobutanol 66, 120
isochlortetracycline 66, 119
iso-elemicin 62
isoprene 71
isothiocyanate, tetrahydrofurfuryl 82
IT-3235 35, 120
IT-3299 35, 119
IT-3353 35, 120
ivy 5
Ixora 4

Japanese persimmon 27
jasmine 4

Kalanchoe 47
kale 17
Karmex 115
kernel 79
kinetin 6, 53, 73, 118
kinins 29
Kling-Tite 23
Krovar 115
Kylar 123

lamb's-quarters 18
larvae 51, 52
–, coccinellid 51
late blight fungus 95
lateral buds 37, 38, 47, 70, 79
latex 55, 69–71, 85, 112, 114, 117, 124
laurylmercaptotetrahydropyrimidine 66, 119
Lavandula angustifolia 6
leaf 10, 19, 21, 25, 27, 33, 34, 37, 45, 52, 55, 77, 81, 91, 93, 94
– drop 21, 25
leafworm, cotton 52
legumes 93, 95

lemon 61, 71, 91, 92
–, rough 6
lentil 95, 98
lettuce 6, 9, 17, 57
light 10, 11
lightwood 72
lima beans 57
lime 61, 91
locust, desert 52
lodge, lodging 33, 34, 42, 43, 54, 93, 112
lung tumor 105
lycopene 60, 61, 71

Macadamia trees 25
magnesium 59
– chlorate 94, 98, 119
mahogany 4
maleic hydrazide 7, 10, 15, 17, 34, 35, 38, 45, 47, 51, 83, 104, 105, 115
malting 32
manganese chlorate 20
marijuana 106
M&B-9057 66, 120
M&B-25-105 49, 122
MBR-6033 35, 67, 124
MBR-12325 35, 65, 116
MCPA 99, 113, 119
mefluidide 34, 35, 65, 68, 79, 116
melon 77, 89
membrane 2
Medok 16, 114
mepiquat 34, 43, 44, 116
β-mercaptovaline 66, 119
merphos 124
metaxon 113, 119
2-(p-methoxybenzyl)-3,4-pyrolidine-diol-3-acetate 66, 119
4-(4-methoxy-6-methylamino-1,3,5-triazine-2-yl)aminocarbonyl benzene sulfonamide 66, 119
2′-methoxy-3-phenacylidene phthalide 101
1-methyl-3-carboxy-4,6-dimethylpyrid-2-one 17, 119
5-methyl-7-chloro-4-ethoxycarbonylmethyl-2,1,3-benzothiadiazole 90, 119
methyl-2-chloro-9-hydroxyfluorene-9-carboxylate 35, 119
2-methyl-4-chlorophenoxyacetic acid 106, 113, 119
methyl decanoate 47, 120
– 3,6-dichloro-o-anisate 66, 120
– 2,7-dichloro-9-hydroxyfluorene-9-carboxylate 35, 120
– N′,N′-dimethyl-N-[(methylcarbamoyl)oxy]-1-thiooximidate 23, 120
2,3:4,6-bis-0-(1-methylethylidene)0-(L)-xylo-2-hexulofuranosonic acid, sodium salt 35, 120
methyl esters of fatty esters 46, 47
– N-ethyldithiocarbanilate 101, 102
– eugenol 62
cis-methyl-eugenol 62
trans-methyl eugenol 62
methyl 9-hydroxyfluorene-9-carboxylate 35, 120
7-methylindole 66, 120
3-(2-methylphenoxy)pyridazine 66, 120
2-methyl-1-propanol 66, 120
methylsulfanil-yl-carbamate 120
metolachlor 99–101
metoxuron 95
metribuzin 101
MH 17, 35, 104, 115
MH-30 115
mice 104, 105
middle lamella 19
millet 16, 57, 101
mineral oil 67
mineral uptake 59
mites, spider, two-spotted 52
Molinate 99, 102
MON-139 36, 122
MON-464 36, 122
MON-814 36, 117
MON-820 35, 122
MON-845 35, 122
MON-8000 67, 122
Monurex 113
monuron 10, 12, 104, 113
morphactins 10, 17, 30, 35, 37, 38, 47, 79, 89
morpholine 33
Motox 111
mulberry 57
mung bean 5
muskmelon 27, 90
Mussaendra 4
mustard 9
–, wild 6
mutagen, mutagenicity 104, 106
mutation 104

NAA 4, 8, 22, 24, 25, 29, 53, 70, 83, 89, 91, 106, 120
NAAm 23–25, 120
NAD 23, 53, 120
naphthaleneacetamide 120
naphthaleneacetic acid (NAA) 4, 8, 22–27, 77, 81, 90, 106, 120
1,8-naphthalic anhydride 99, 101, 102, 121
2-naphthoxyacetic acid (NOA) 28, 29, 121
1-naphthyl-N-methylcarbamate 23, 24, 121

N-1-naphthylphthalamic acid 25, 121
naptalam 52, 121
Naptro 98
NaTA 124
naval stores 72
navel orange 28, 58, 75, 91
navy beans 57
nematicide 23, 120
Niagara Stik 120
nicotine 62
nitroanilines 46
nitrogen 59, 62, 68
nitrophenylhydrazines 46
NOA 53, 121
node 9, 34, 40, 41
NPA-3 121
nutrient uptake 33
nuts 78
–, cashew 29
–, *Macadamia* 25
–, pea- 79, 110
–, pistachio 78
Nyctanthes 4

oats 3, 57, 95, 99, 101
–, wild 6, 99
octamethylenediamine 101
oil 108
–, essential 61
–, mineral 67
–, palm 89
–, peanut 79
–, tall 9
okra 59
Oleander 47
oleoresin 72, 116
olive 22
onion 7
orange 21, 22, 25, 27, 28, 60, 61, 76, 91
–, navel 28, 58, 75, 91
–, sour 6
–, sweet 6
–, trifoliate 6
organ size 32
organohalides 105
Oriental persimmon 28
Orthene 115
Ortho-12420 115
Ortran 115
ovule 28
7-oxabicyclo(2:2:1)heptane-2,3-dicarboxylic acid 35, 66, 98, 121
oxamyl 23, 24, 120

Pachystachis 47
palm, oil 89
papain 70
papaya 70
paprika pepper 80
paraquat 21, 72, 94–97, 101, 104–106, 116
parasitism 51
Parthenium argentatum 71
PBA 50, 111
PCP 121
pea 5, 6, 32, 34, 55
– aphid 51
peach 4, 6, 9, 25, 28, 73, 78, 79
peanuts 57, 59, 110
pear 9, 28, 30, 49, 80
pecans 47
pectinase 19
peel 21, 60, 76
pegs, peanut 79
Pen-V 66, 121
Pen-Vee 66, 121
penicillamine 66, 119
penicillin V 66, 121
penta 121
pentachlorophenate, sodium 94, 121
n-pentanoic acid 66, 121
Peperomia 4
pepper 17, 56, 80
–, banana 80
–, bell 61
–, chili 61
–, paprika 80
–, pimiento 61, 80
–, red 80
persimmon, Japanese 27
persimmon, Oriental 28
pesticide 1, 104, 107
petal 30
Phenacide 111
6-phenoxyacetamido-penicillanic acid 66, 121
phenoxyacetic acid 30, 121
phenoxy acids 105
N-(2-phenoxyethyl)-N-propyl-1H-imidazole-1-carboxamide 66, 121
2-(1-phenylethylsulfonyl)pyridine-N-oxide 36, 121
phenylglyoxylonitrile-2-oximecyanomethylether 101, 102
phenylmercuric acetate 55, 56, 121
1-(phenylmethyl)-1H-pyrrole-2,3-dicarboxylic acid 101
N-phenylphosphinylmethyliminodiacetic acid-N-oxide 67, 121
N-phenylsulfonamido-N-phosphonomethylglycine 67, 121
3-phenyl-1,2,4-thiadiazol-5-yl-thioacetic acid 49, 121
N-phenyl-N′-1,2,3-thiadiazol-5-yl urea 20, 121

Phosphon 55, 57, 123
Phosphon D 6
Phosphon S 55
phosphonic acid, (2,2,2-trichloro-1-hydroxyethyl)-bis-[(2-hydroxypropoxy)-1-methylethyl]ester 67, 122
phosphonomethylglycine 36, 67, 68, 97, 122
phosphonomethylglycine, calcium salt 36, 122
N,N-bis-(phosphonomethyl)glycine 35, 68, 122
N-phosphonomethyliminodiacetic acid 35, 122
phosphorus 59
photosynthesis 56, 99
Phytar-138 65
Phytar-560 116, 118
phytotoxicity 18, 21, 25, 34, 68, 91, 99
pickling cucumbers 29
picloram 64, 79, 110
Pieris 47
pigment 60, 61, 71, 75, 76, 80, 112–114, 123
pigweed, red root 18
Pik-Off 22, 117
pimiento pepper 61, 80
Pinaceae 14
pinching, chemical 47
pine 4, 56, 72
pineapple 3, 8, 9, 28, 88
pink bollworm, cotton 52, 53
Pinus taeda 14
piproctanylium bromide 34
pistachio nuts 78
pit 73
Pix 116
plant composition 60
– shape 49, 107
– size 32, 107
Plantgard 114
Plucker 23
plum 4, 25, 28, 49, 80, 81
PMA 121
POA 121
pod 79
– set 81
Poinsettia 5, 9, 34, 48, 57
Polado 67, 68, 122
Polaris 35, 64, 68, 122
pole bean 57
pollen 15, 28, 30, 33
pollination 90
pollution, environmental 69
polyamines 57
polychlorocarboxylic acid esters 17, 122
poly[oxyethylene(dimethylimino)ethylene-(dimethylimino)-ethylene-dichloride] 67, 122
polyphenoloxidase 79
pome fruit 72
pomologist 72
poplar 4, 5
Po-San 119
potassium 59
potassium 3,4-dichloro-isothiazole-5-carboxylate 21, 122
potato 7, 93–95, 97, 104
– scab, common 54
PP-333 35
PP-757 66, 120
Premerge 3 98, 111, 117
primings, tobacco 82
Pro-Gibb 125
Promalin 30, 31, 73
propagation 4
N-(2′-n-propoxyethyl)-2,6-diethylchloroacetanilide 102
propyl-3-t-butylphenoxy acetate 49, 122
Protea 4
protein 59
proteolytic enzyme 70
prune 80
pruning, chemical 47–49, 118, 120
pumpkin 60, 89
pupae 51, 52
Pyracantha 47
pyridazinones 46

quality 1, 63, 72, 74, 77, 92
quiescence 6

R-25788 99, 114
rabbiteye blueberry 74
Racuza 66, 120
radish 9, 82
ragweed 18
rape plant 33
raspberry 37, 57, 74
rats 104–106
ratoon 50, 68
red pepper 80
redroot pigweed 18
Regime-8 125
Reglone 97, 115
Reglox 97, 115
Re-Green 97
regulations 20
Release 21, 22, 113
resistance 51
retardants 6, 8, 9, 29, 34–38, 51, 52, 54, 55, 57, 82, 110–112, 114–117, 119–124
RH-531 16, 113
RH-532 17

rhizobitoxine analogs 91
Rhododendron 4, 12, 47
Rhonox 113, 119
rice 32, 33, 50, 57, 93, 94, 97–102
rind 58, 75, 76, 87, 91
ripeners, ripening 9, 62–69, 72–76, 78, 80–84, 86, 87, 95, 110–125
Ripenthol 66, 68, 69, 121
root 4, 55, 71
– formation 3–5, 70, 114, 118, 120, 124
rose 4, 13, 91
rot, charcoal 53
rough lemon 6
Round-Up 36, 68, 97, 122
RPAR 104
rubber 4, 69–71, 85, 108
– tree 69–71, 85
runner bean 5
rye 16, 17
ryegrass 16

SADH 123
safety, environmental 104
–, human 104
salicylaldoxime 56, 122
salinity 55
salt 57
– tolerance 57
scab, potato, common 54
scald, apple 72
seed 6, 7, 14, 15, 28, 57, 79, 81, 89, 90, 96, 99
– coat 7
– head 34
seedless grapes 78, 125
seedling 3, 32–34, 57, 77
seed pieces 50
senescence 52, 76, 91, 111, 112, 114, 120, 123, 125
setts, sugarcane 50
Sevin 23, 24, 121
sex change(r) 89, 110–112, 117, 119, 120, 125
shape, fruit ` 30, 31
–, plant 49
shelf life 91
side branches 32, 47
silage 77
silicone 55
silver nitrate 90
Sinox 23, 117
Sitka spruce 14
size, organ 32
–, plant 32
Slo-Gro 35
smoke 8
snap beans 52
snapdragon 91
sod-forming-grass 50
sodium chlorate 20, 94, 97, 105, 122
– chloroacetate 94, 98, 122
sodium-1-(p-chlorophenyl)-1,2-dihydro-4,6-dimethyl-2-oxonicotinate 16, 82
sodium p-(dimethylamino)benzenediazosulfonate 101
– 4,6-dinitro-o-cresylate 23, 24
– p-methylbenzenediazosulfonate 101
– pentachlorophenate 94, 98, 121
– polysulfide 22
solvent 120
sorghum 6, 56, 99, 101, 102
sorghum, grain 94, 95, 97, 99–101
sour cherry 75
– orange 6
southern bean blight 15
soybean 16, 53, 57, 59, 81, 82, 91, 93–95, 97–101, 113, 123, 125
Spark 111, 117
spider mites, two-spotted 52
spinach 57
spindle 10
spreaders 20
Sprout-Nip 112
sprouting 7
Sprout Stop 115
spruce, Sitka 14
Spud-Nic 112
squash 28, 89
–, zuccini 57
Stafast 23
stem 19, 34, 37, 43, 59, 71
sterility 15, 18
steroid 33
Stevia rebaudiana 13
stevioside 13
stickers 20
stickiness, rind 76
stinkweed 6
stomate 55, 56
stone fruit 9, 25, 28, 73
strawberry 56
stress, environmental 55, 108
stripper harvest 93
stumps 70
succinic acid 2,2-dimethylhydrazine 123
sucker 38, 45, 46, 77
sucker-stuff 115
sucrose 10, 63, 91
sugar 10, 62, 69, 70, 72, 75, 78
sugarbeet 16
sugarcane 9–12, 32, 40, 41, 50, 51, 63–69, 77, 78, 83, 84, 96, 97, 108, 110–116, 118–124
– borer 51

sugar maple 56
sulfuric acid 95
sunflower 17, 47, 56, 95, 97
surfactants 20, 48
Sustar 35, 67, 124
Sweep 22
sweet cherries 25, 61, 75
– orange 6

2,4,5-T 4, 25, 70, 73, 78, 79, 83, 107, 124
2,4,6-T 99
tall oil trees 9
tangerine tomato 71
tapping 70, 85
tasseling 10
Taxodiaceae 14
TBA 67, 124
TCA 98, 124
TD-191 66
tea 4
Telvar 113
temperature 8–10, 30, 55, 57, 72, 108
teratogen 106
terbutryn 101
Tergitol NPX 67
Terstroemia 47
tetrahydrobenzoic acid 65, 113
tetrahydrofurfuryl isothiocyanate 82, 123
tetrahydrofuroic acid hydrazide 67, 123
(22R,23R,24S)-2α-3α,22,23-tetrahydroxy-24-methyl-6,7-s-5α-cholestano-6,7-lactone 123
1,1,5,5-tetramethyl-3-dimethylamino-dithiobiuret 25, 123
TH-6241 90, 119
thinning 22–26, 28, 111–113, 117, 120, 121, 123
thiocarbamates 99, 100
thistle, Canada 103
TIBA 17, 50, 53, 79, 81, 89, 125
tillers, tillering 29, 50, 111, 112, 119, 125
Tip-Off 23
tobacco 16, 38, 45, 46, 54, 56, 57, 62, 82
– budworm 53
n-m-tolylphthalamic acid 29, 123
tomato 5, 15–17, 28, 30, 47, 53, 56, 57, 59, 61, 71, 82, 86, 87, 99
topping 37
Tordon 64, 110
toxaphene 11, 111
toxicity, toxicology 20, 45, 52, 56, 57, 100, 104–106
2,4,5-TP 73, 124
translocation 59, 103, 119
transpiration 55, 72
Transplantone 120
tree 73, 75, 80
–, apple 49, 52, 73, 83
– bark 70
–, cherry 49
–, cherry, sweet 49
–, coffee 27
–, forest 4, 14
–, fruit 49, 57
–, pear 49
–, pecan 47
–, plum 49
–, rubber 69–71, 85
–, tulip 4
–, walnut 47
trefoil 96, 97
Tre-Hold 23, 120
triacontanol 33, 77, 118, 123
triallate 99, 102
triazines 100
1,2,4-triazine-3,5(2H,4H)-dione 67, 123
tributyl(5-chloro-2-thienyl)methyl-phosphonium chloride 54, 82, 105, 123
tributyl-2,4-(dichlorobenzyl)phosphonium chloride 123
S,S,S-tributylphosphorotrithioate 20, 97, 105, 124
tributylphosphorotrithioite 20, 124
trichloroacetic acid 94, 98, 124
N-trichloroacetylaminomethylene-phosphonic acid 67, 124
5-(2,3,3-trichloroallyl)-N,N-diisothio-carbamate 102
2,3,6-trichlorobenzoic acid 51, 63, 67, 104, 124
2,4,5-trichlorophenoxyacetic acid 4, 77, 124
2,4,6-trichlorophenoxyacetic acid 99, 124
2,4,5-trichlorophenoxypropionic acid 77, 124
triethylamines 61, 71
bis(N,O-trifluoroacetyl)-N-phosphono-methylglycine 64, 124
2-trifluoromethylquinoline carboxylic acid 57
3-trifluoromethylsulfonamido-p-acetotoluidide 35, 67, 124
trifoliate orange 6
2,4a,7-trihydroxy-1-methyl-8-methylene gibb-3-ene-1,10-dicarboxylic-acid-1,4-lactone 17, 125
2,3,5-triiodobenzoic acid 17, 50, 125
triticale 16, 17
Trysben 67, 124
tuber 7
tufted grasses 50
tulip tree 4
tumbleleaf 97

tumor agent, anti- 46
–, lung 105
turf grass 34
Tween-20 67
typey fruit 30

undecanol 48
UNI-D513 17
UNI-P293 35
uptake, nutrient 33, 59
urethane 105
Urox 115

n-valeric acid 66, 121
vanillin 66, 78, 118
Varitox 98, 124
vase life 91
vegetables 9, 32, 71, 81–83
Verticillium wilt 53, 54
vessels, latex 70
vines, grape 26, 27, 37
–, peanut 79
–, potato 95
Viola odorata 6
virus 53, 54
Vydate 23, 24, 120

walnut 25, 47
water 10, 55
– core 72
wax bean 52
weed 93–103, 107
Weedar 113, 119, 124
Weedone 124
Weedbeads 98
weed control 51, 99, 103, 107
Weed-B-Gon 114
weevil, boll 53
wetting agents 20
wheat 16, 17, 29, 33, 34, 43, 50, 54, 57, 59, 95, 99–102
wild garlic 103
– mustard 6
– oats 6, 99
wilt, *Fusarium* 53
–, *Verticillium* 53, 54
winged bean 4
wood 72
– preservative 121
worm, leaf-, cotton 52

Xanthium strumarium 56
xylem 72
Xylosma 47

yemane 4
yield 1, 9, 10, 22, 27, 28, 32–34, 37, 55, 63, 70, 71, 74, 76, 77, 79, 82, 89, 94, 96, 111, 117

Zebrina pendula 5
zinc sulfate 22
Zinnia 47
Zizyphus mauritiana 90
Zotox 97, 110
zoysia grass 42
zuccini squash 57

MIX
Papier aus verantwortungsvollen Quellen
Paper from responsible sources
FSC® C105338

If you have any concerns about our products,
you can contact us on
ProductSafety@springernature.com

In case Publisher is established outside the EU,
the EU authorized representative is:
Springer Nature Customer Service Center GmbH
Europaplatz 3, 69115 Heidelberg, Germany

Printed by Libri Plureos GmbH
in Hamburg, Germany